Waters of the Deep

A Biblical and Scientific Exploration of Creation and the Flood

Jonathan Freed, MD

Carpenter's Son Publishing

Waters of the Deep: A Biblical and Scientific Exploration
of Creation and the Flood

Published by Carpenter's Son Publishing, Franklin, Tennessee
www.ClovercroftPublishingGroup.com

Cover and Interior Design by Suzanne Lawing

ISBN: 978-1-968127-28-2 (print)

Printed in the United States of America

Dedication

To my father
who passed a quest for knowledge
of God and science
on to me

Contents

INTRODUCTION

In the beginning God created the heavens and the earth.
— Genesis 1:1

Few verses in the Bible have generated more discussion, debate, and interpretation than this one. For thousands of years, believers have sought to understand what exactly the words "in the beginning" meant and how God's creative acts unfolded. Were the heavens and the earth created together? Did God form the stars and galaxies at the same time He shaped the world we inhabit? And how does the scientific record—the fossil layers, the cosmic background radiation, the vastness of the universe—fit into the biblical story of creation?

For centuries, thinkers have wrestled with the intersection of Scripture and science. The debate has not merely been about physical origins, but about the very nature of God, His Word, and the reliability of divine revelation. Today, Christians find themselves confronted by modern cosmology and evolutionary theory, both of which often appear to contradict the Genesis account. Yet millions of thoughtful believers are convinced that true science, rightly understood, harmonizes with Scripture rather than undermines it.

The Tension Between Faith and Science

The rise of modern science brought with it extraordinary discoveries. We now peer into galaxies billions of light-years away and split the atom into subatomic particles. We trace DNA to its molecular coding and chart the history of our planet through its geologic layers. But alongside these advances came a philosophical shift—the conviction that all phenomena could be explained by natural causes alone, without reference to God.

This assumption, known as naturalism, has shaped much of contemporary science. If the cosmos is self-contained and self-explanatory, then divine creation must be reinterpreted—or rejected altogether. Yet this worldview, while dominant in academia, is not itself scientific; it is a philosophical choice, one that excludes the supernatural before examining the evidence.

Christians who accept the authority of Scripture face a challenge: how do we interpret the opening chapters of Genesis in light of scientific findings without compromising biblical truth? Should we adjust our theology to fit current models of cosmology and geology, or should we let the Bible interpret the data for us?

The Importance of Genesis

Genesis is not a peripheral text. It forms the foundation of the entire biblical narrative—explaining who God is, who we are, and why the world is the way it is. The New Testament writers, and even Jesus Himself, repeatedly refer back to Genesis as literal history. The doctrines of sin, redemption, the Sabbath, marriage, and salvation are all rooted in the early chapters of Scripture.

If Genesis is mere allegory, then the gospel loses its historical grounding. If Adam and Eve were not real people, then the Fall is symbolic, and Christ's death and resurrection—intended to reverse that Fall—become theological metaphors rather than historical events. Thus, our understanding of creation directly affects our understanding of salvation.

Competing Models

Over the past two centuries, numerous theories have been proposed to reconcile biblical revelation with scientific discovery. Some Christians have adopted Theistic Evolution, arguing that God used natural evolutionary processes to bring about life over billions of years. Others accept an Old Earth but reject evolution, proposing that the universe and our planet developed naturally over eons before God created life. Many others hold to a Young Earth view, asserting that all creation occurred in six literal days only a few thousand years ago.

Still others, seeking harmony between the biblical and scientific records, suggest hybrid models—such as the Gap Theory, Day-Age Theory, or Framework Hypothesis—each attempting to bridge perceived contradictions between Scripture and science.

These competing models reflect different hermeneutical approaches to Genesis. Some take the text as strictly literal, others as partly figurative, and still others as primarily theological. The challenge lies in finding a model that both respects the authority of Scripture and coherently integrates observable evidence from the natural world.

Why It Matters

Some argue that such debates are secondary, that our faith does not depend on the details of creation. Yet this issue shapes our worldview at its core. What we believe about origins influences how we view humanity, morality, purpose, and destiny. If we are the result of divine design, then life has meaning and accountability. If we are cosmic accidents, the product of blind chance, then morality and purpose are illusions.

Moreover, the creation account reveals God's character. A universe created by random processes would reflect randomness and indifference. A universe spoken into existence by a personal, purposeful Creator reflects intelligence, beauty, and love.

The Aim of This Book

Waters of the Deep seeks to explore creation through both the lens of Scripture and the insights of science. We will examine the evidence for a Hydroplate model of the flood, a framework developed by creation scientist Dr. Walter Brown and adapted here to harmonize with an Old Universe / Young Earth understanding of origins. This model provides a coherent explanation of the global flood, geological formations, and the chronology of Genesis.

The goal is not to pit science against Scripture, but to show that when both are properly understood, they speak with one voice. The Bible's authors were not primitive or naïve. They wrote under divine inspiration, describing reality in language suited to their time yet profoundly consistent with truths still being discovered today.

In the chapters ahead, we will,

- Define and compare the major creation models;
- Examine the Hydroplate Theory in depth;
- Explore how this theory aligns with Scripture and known science;
- Consider the theological and spiritual implications of creation and the flood.

As we proceed, we will find that faith and reason are not adversaries but partners in the search for truth. The God who inspired Scripture also created the cosmos; therefore, the two cannot ultimately contradict each other.

A Call to Humility and Awe

Whether one peers through a telescope at the stars or through a microscope at the cell, the same conclusion emerges: Creation is astonishingly complex and exquisitely ordered. The more we learn, the more we realize how little we truly understand. As Job confessed after encountering the Creator face-to-face:

> *I have uttered what I did not understand, things too wonderful for me, which I did not know.*
>
> — Job 42:3

The study of origins should therefore lead not to pride or contention, but to humility and worship. The question is not simply how God created, but who this God is—and how we, His creation, should respond to Him.

CHAPTER 1

Conflicting Creation Models

As you do not know what is the way of the wind, or how the bones grow in the womb of her who is with child, so you do not know the works of God who makes everything.

— ECCLESIASTES 11:5

My wife and I share warm memories of Thanksgivings with our sons and their wives in our house as the fireplace crackles and we gather around a large table. Per ritual, year after year, we piece together a large jigsaw puzzle. The project takes days, and though we know what the final outcome will be, the details of how the picture will fit together remain uncertain. Some in the family will diligently focus hours of attention on the creation; others will dabble with waxing and waning interest. But the end result, however long it takes and regardless of the steps taken to piece it together, is always consistent with the beautiful picture on the puzzle box.

The creation process of the puzzle often begins with the straight edges forming an outline—the borders, in fact, of what is to come. And from there, it progresses to similar colors being placed together, or separated, depending upon the picture before us. Portions of the whole are, at times, created and collected on the sides, to be added later; others are added to the major framework right away. It's not always easy to know what goes where, and the puzzle leaves no room for error. That is, no matter how closely matched the colors, or even the prospective shapes, the wrong piece will not fit—and cannot (no matter how at times attempted) be forced into place.

The study of our creation is similar to the process of putting together the puzzle. Humans, from the beginning of recorded history, have wondered about their origins, as the numerous ancient creation accounts have revealed. After all, we are here, and nothing about us or the creation specifically tells us about how we got here, or why. Its grandeur and awesomeness point to a Creator or creators; at least that is what the majority of humanity had believed for most of history. In modernity, with the advent of the French Revolution and the Age of Enlightenment, the idea that science alone—without the help of a transcendent God—can account for creation has taken hold. Mindless forces and matter alone are left to account for human origins.

What is the answer? Like the puzzle before my family each Thanksgiving, we know what the result is: our created world. And, like the puzzle, there are numerous pieces that we must try to fit together in order to create a logical thesis that explains the existence of that created world. And as with the puzzle, we must recognize when aspects of our model do not match the data or the reality before us. At times, the theory fails because it isn't consistent with the whole picture; or it logically, rationally, or scientifically fails. However, when we have all the correct pieces in all the right places, they work together nicely and form the beautiful picture on the box. Our task is to ensure that we have all the correct pieces in all the correct locations. In other words, to differentiate between what works and what doesn't.

What follows, then, are *some* of the biggest, broadest creation models presented today. Though each comes in various forms, they seek to answer the age-old question, perhaps the most important question of all: What is our origin? In other words, who are we, why are we here, and ultimately, where are we going?

Secular Cosmology: Abiogenesis

Every age has its own creation story, its own account of human origins. From the Egypt of the Pharaohs to the America of Richard Dawkins, these stories reflect, in some way, the age and culture from which they arise. Brewing up from the present scientific materialist worldview, one creation account—abiogenesis (life arose from inorganic matter without anything supernatural)—goes roughly like this: About 13.8 billion years ago, an event called (at first derisively, when coined in 1949) the "Big Bang"[1] unfolded. According to this model, there was "a singularity," an infinitely hot, infinitely dense something or other where all the known laws of physics broke down and didn't apply. They have no workable theory as to the origins of this singularity, and they can't talk about what came "before" it because, according to the theory, time itself had not yet existed.[2]

Through a series of events described, possibly by quantum field theory (a branch of physics developed using special relativity and quantum mechanics to try to account for the universe's structure), this singularity exploded—creating space, time, matter, and energy. Subatomic particles, spawned in the Big Bang, congealed into protons, neutrons, and electrons. As the protons and neutrons moved through the newly created space, they cooled and began to form hydrogen and helium atoms, which composed most of the early universe.

Then gravitational attraction, one of the four fundamental forces in the universe, brought these elements together to form protostars. Some of these stars gathered enough energy to become massive fusion reactors—factories squeezing smaller elements together into larger atomic nuclei. Heavier elements were synthesized forming carbon, oxygen, and nitrogen. These elements continued to condense into the galaxies and star formations that exist today. The key to everything is gravity, described by Einstein's General Relativity.

However, even our best theories cannot explain aspects of the universe according to General Relativity, because the observable mass and its expected gravitational force does not account for the shape and motion of the surrounding galaxies, causing some cosmologists to speculate that most of the universe is composed of something called "dark matter" and "dark energy."[3] These unobservable factors (called "dark" for this reason) are the only way so far to explain observable phenomena in the universe, such as its increasing speed of expansion, at least according to General Relativity.

While the Big Bang model remains dominant, several physicists—including Eric Lerner and Jayant Narlikar—have proposed alternative cosmologies, and even proponents of the standard model have acknowledged unresolved problems within its framework.[4]

Nevertheless, according to the Big Bang model, about 4.6 billion years ago our sun and earth condensed from this space debris; as a result, the solar system began to form. As these massive globules cooled down, oxygen (supposedly made by the stars too) and hydrogen combined to form water—in outer space itself! This water floating through outer space (very hard to envision), along with heavier elements, including nickel and iron, which make up the core of the earth, condensed on asteroids. Eventually, these asteroids bombarded the earth, providing it with water, nickel, iron, and many of the other elements here now. Then, the speculation goes, those heavier elements sank to earth's core in the initially molten earth. The lighter elements stayed near the surface, although for some unknown reason, some of the heavier elements, such as iron, stayed near earth's surface as well.[5]

However, geologists have recently begun to question this theory, called "Extra-terrestrial Delivery," because they doubt that enough water could have been delivered to the earth by asteroids in order to account for the oceans.[6] There is, after all, a lot of water here! Depending on assumptions about impact retention efficiency, water content of asteroidal material, and whether earth's interior water is included, asteroid-delivery models can require an incoming mass comparable to a Mars-sized body or greater.[7] Even more problematic is the heat that this mass of asteroids would have created on impact: It would have easily vaporized all of the water that these asteroids supposedly brought here to begin with.[8] However popular the theory, it sounds very far-fetched, like a drastic leap of faith, actually.

The next problem for the secular theory of evolution to tackle was how life could have evolved on this planet called earth. The theory moves on to postulate how conditions for such an event could have arisen.

Astronomers suggest that our sun is in the middle of its ten-billion-year life cycle, about five billion years old.[9] The early sun would have been relatively "dim," but with increasing intensity it would have provided the earth with increasing warmth. This idea poses a problem, though. The temperature on early earth, perhaps about 3.8 billion years ago (when the sun was only 1.2 billion years old), would have created a frozen, sterile wilderness utterly unsuitable for life. Scientists have, therefore, postulated a greenhouse gas effect that would have kept the temperature "just so." This idea is referred to as the "Goldilocks Effect"—an earth not too hot, not too cold. If the earth were colder, no life could have existed on the frigid planet; if warmer, all its water would have evaporated, which meant no life, either. So, as the earth warmed, the theory goes, the greenhouse effect helped the temperature to stay uniform, and just right for the development of life.[10] That, at least, is the speculation.

Also, assuming the temperature was right (a big assumption)—life, according to one model, is thought to have developed about 3.8 billion years ago on earth from a primordial soup of chemicals (though the "primordial soup" theory as well has long been challenged even by secular scientists).[11] This process required the building blocks of life to be present, obviously. More complex molecules had to develop and then accumulate in just the right concentrations, an incredibly complex balance that presents numerous challenges to the whole idea of the abiogenesis theory of chemical evolution.

First, chemicals do not spontaneously tend toward greater complexity, much less to spontaneously form life—not even close. Chemicals tend to decay, not to increase in complexity, especially the complexity needed for even the "simplest life" (a phrase that becomes more oxymoronic as our knowledge of the cell each year increases). The "natural selection" process that forms the background to evolutionary theory does not apply to lifeless chemicals. Chemicals have no preference for life or non-life. There is no "natural selection" with them. Abiogenesis requires just the right concentration of reagents, in just the right place, with just the right conditions. Then, once life spontaneously formed, there had to be some mechanism to sustain and preserve this life. The odds of molecules having randomly assembled into a living organism, given the proposed age of the universe itself, are so astronomically improbable that one hard-core atheist, Francis Crick (he of DNA fame) believed life must have come from somewhere else in the universe first.[12] That is, life could not have formed by chance, and so one of the greatest scientific minds of the last century, a Nobel Prize winner to boot, had extra-terrestrials coming down in rocket ships and seeding life here.

The situation gets only worse for abiogenesis. Along with life itself developing by chance alone (stunningly unlikely) was the requirement for the information, exceedingly complex and refined information, needed to replicate the same patterns of chemicals and their behaviors in order to make new copies of the earliest life form. Knowing the deep problem, some biologists have suggested something called the "protocell," an early cell-like structure with rudimentary membranes and organelles, using mainly RNA to code for replication and protein synthesis[13]—an exceedingly weak and highly speculative ad hoc attempt to explain what remains inexplicable.

Then, some of these abiogenic processes required oxygen and water, which is believed to have been there, in abundance. However, other aspects of chemical evolutionary theory require that there *be no oxygen and no water present.*[14] Oxygen or no oxygen, water or no water, the theory teaches that this early life, however it arose, slowly developed through a series of random mutations into more and more complex entities. Yet the chances of these mutations creating increased complexity to the point where life was self-perpetuating was less than 1:1080, the estimated number of atoms in the universe.[15] Not only is this rise in complexity unlikely to the point of absurdity, the preservation of these new life forms would have been staggeringly difficult as well. The earth was an ecosystem with incredible instability over time. There would have been continual variations in temperature, water content, the atmosphere and oxygenation—a reality very hostile to life itself.

Regardless of these vast leaps of speculative faith and ad hoc stories buttressing the abiogenic theory of life's origins and complexity, we are then to believe that this life slowly evolved into multicellular organisms that resulted in the Cambrian Explosion, supposedly about 540 million years ago.[16,17] This Cambrian Explosion refers to the sudden appearance of multiple life forms in the lowest layers of sedimentary rock, with no transitory forms below. It just seemed to show up, out of nowhere, which for years has challenged the evolutionary model, leading to more ad hoc speculation to try and explain how these fossils, some of them very complicated organisms, could have come into existence when, according to mainstream evolutionary theory, they should not have.

Besides all the creatures found in the Cambrian level, these early protocells supposedly evolved into, among other things, dinosaurs—including some weighing 125,000 pounds (quite a transition from a tiny cell to a Brontosaurus)—about 250 million years ago. This allows just under 300 million years for development of animals from single cells to this massive 125,000-pound size. The beasts roamed the earth until about 66 million years ago, when an asteroid (them things again!) struck the earth, the equivalent to billions of nuclear bombs, and, filling the atmosphere with so much debris, the asteroid created a "nuclear winter" scenario that killed off the dinosaurs.[18] However, other life forms somehow survived and continued to evolve.

Meanwhile, there are the ice ages, too, five distinct ones to be exact.[19] The earliest one occurred 2.4 billion years ago. The second occurred about 700 million years ago and covered the entire earth's surface. (The dinosaurs would have emerged a short 100 million years after this ice age.) There were, we're told, two more ice ages until the most recent one, a glacial epoch developing just 2.6 million years ago, peaking 20,000 years ago—with us now on the downhill side of a thawing-out phase. Thus, we have faced global warming long before the advent of the combustion engine; that is, at least according to the theory.

Also, according to the theory, as life continued to develop and evolve in complexity, hominids emerged about six million years ago.[20] The first fully human species appeared around 300,000 years ago, supposedly. Most mainstream anthropologists believe that civilizations developed out of Africa. However, more recent research is finding evidence for civilizations emerging from Asia Minor and modern-day Turkey. Art, language, culture, even religion developed about 50,000 to 70,000 years ago. The earliest documented piece of human writing is the Kish Tablet, circa 3500 BC, from the regions of Mesopotamia (modern-day Iraq).

How fascinating, if not astonishingly "coincidental," that some of the earliest known aspects of human civilization appear in the region where Noah's ark is believed to have landed. Also interesting is how we have no traces of human writing until just after the estimated time of the flood when, according to the Bible, the survivors began to spread out and multiply.

Though there are vast differences among those who promote this model—the leading voices being personages like Richard Dawkins, Neil deGrasse Tyson, Steven Pinker, and others—all hold passionately, dogmatically (even fanatically) to one point that they must take on faith: denial of any supernatural or divine intervention in the entire process. From the origin of the "singularity," to the unfathomable complexity of the human brain, everything was the result of billions of years of an unconscious and undirected confluence of fortuitous events that just happened to create the universe and us here on it.

Most insightful is evolutionary biologist Richard Lewontin's famous quote in the *New York Review of Books* (January 9, 1997):

> We take the side of science in spite of the patent absurdity of some of its constructs, in spite of its failure to fulfill many of its extravagant promises of health and life, in spite of the tolerance of the scientific community for unsubstantiated just-so stories, because we have a prior commitment, a commitment to materialism. It is not that the methods and institutions of science somehow compel us to accept a material explanation of the phenomenal world, but, on the contrary, that we are forced by our a priori adherence to material causes to create an apparatus of investigation and a set of concepts that produce material explanations, no matter how counter-intuitive, no matter how mystifying to the uninitiated. Moreover, that materialism is absolute, for we cannot allow a Divine Foot in the door.[21]

Some, however, want a divine foot in the door, and how they let that foot in, and what it means when they do, is what the next section deals with.

Theistic Evolution

For many Christians, atheist evolution is a non-starter, an illogical, irrational—and ideologically driven (see above)—attempt to explain creation, but without the moral imperatives implied by the existence of a Creator. As with its atheistic incantations, theistic evolution comes in many forms, though all theistic evolutionists accept that, if there really were a Big Bang, then God Himself had done it (as some quip: "God said 'Bang!' and there it was"). This option certainly makes more sense than just assuming the "singularity"—from which the entire creation arose—as a given, with no explanation whatsoever. The universe, as one cosmologist said, *is just one of those things that happen from time to time,* about as evasive an "answer" for origins as possible without flat-out admitting that you don't know.

For some theistic evolutionists, God was the great and almighty Initiator—and that's it. Basically, theirs is the god of deism: He (She? It?) got the cosmos started, and then absconded, indifferent to our fate. This God set in motion—and then completely washed his hands of them (and us)—the factors that allowed the evolution of humanity after billions of years.

Others view the Creator as more involved than was this *Deus absconditus.* Yes, He created the world in a way that would allow it to evolve, and perhaps at various times He even interceded (or perhaps not?) to make sure that things just didn't die out completely. Whatever God's continued role (or lack thereof), after eons of the dog-eat-dog-survival-of-the-fittest struggle (pretty much what secular evolution teaches)—two protohumans, "Adam" and "Eve," evolved enough for God to then imbue them with His "spirit." Hence, they became self-conscious rational beings with intelligence, a moral compass, and the ability to think abstractly—pretty much what helps separate humans from all other living creatures here on earth now. Thus, with the arrival of late hominid monkeys, having now been given a divine "spark" or "spirit" or whatever, the Bible story begins.

From this theistic evolutionary perspective, the Bible, especially the first twelve chapters, is deemed as "mytho-history," basically stories that didn't literally happen (such as a real serpent in a real Eden deceiving a real Eve) but that, nevertheless, teach important spiritual truths or the like. Knowing how irrational the idea that all the beauty, design, and overt intentionality found in nature just happened by chance, or that it all arose from nothing, they nevertheless feel impelled to believe what secular science teaches about human origins, and so they interpret Scripture through the lens of evolution. This, however, is not an easy fit (to say the least), and it comes heavy-laden with implications to which, perhaps, many theistic evolutionists have not given enough thought. (See Chapter 5.)

Nevertheless, some highly educated and intelligent people—such as Francis Collins,[22] Kenneth Miller,[23] Alister McGrath,[24] and William Lane Craig[25] et al.—have accepted some version of theistic evolution. That is, though "believing" in the Bible, they try to harmonize it with modern science's present theory about origins.

Active Gap Theory

Some theistic evolutionists believe in what has been called by various names: the "Active Gap Theory," the "Pre-Adamite Theory," the "First Earth Doctrine," and the "Ruin-Restoration Hypothesis." However named, and whatever the variations promulgated among theorists (the few remaining), it goes roughly like this:

Sometime in the distant past, perhaps billions of years ago, God created the heavens and the earth. That is, He created the entire cosmos: all the stars and galaxies we see from a great distance, and everything we see here on planet earth. Life, animals, plants, all that we see today was created in Genesis 1:1. After this, there was an "active gap" of time where things evolved and changed.

In the two views looked at above, there was either no God, or, if there was one, then outside of initiating things, He was far removed from the creation. In contrast, according to this pre-Adamite theory, God was intimately involved with His creation at every step. Nothing, from star dust to the Andromeda Galaxy, and everything in between, natural and supernatural, was left to chance, including events on this earth.

And, indeed, for billions of years, things went quite well, actually. The earth itself was filled with life thriving in a perfect environment. Various theories speculate about the nature of that life, particularly about "pre-Adamites," humans (or human-like beings) existing before the biblical Adam, who, in the Abrahamic religions, is traditionally considered the first man. This notion of these pre-Adamites had appeared in religious, philosophical, and scientific discussions over the centuries, in attempts to reconcile emerging scientific knowledge with theological beliefs.

Who were these pre-Adamites? Some saw them as primitive, or animal-like, kind of proto-but soulless humans. Others saw them as pure spirit beings who "fell" into material form (a very pagan Greek concept). French philosopher Isaac La Peyrère (1596–1676) argued that Adam was the ancestor of the Jews only, and that other peoples, the Gentiles, descended from the pre-Adamites. Others argued that the pre-Adamites were giants, spiritually enlightened beings who, ultimately, degenerated into more earthly material entities.

Whatever their nature, the pristine world that they lived in was, for some reason (perhaps caused by Satan and his angels?) shattered—and catastrophe enveloped the earth. This destruction of the earth brought eons of suffering and death, which lead to the development of the paleontological strata and record embedded in the earth today. The earth during this long time became "formless and void." God, then, returned to this devastated planet and recreated it, placing Adam and Eve in the Garden of Eden—and thus began biblical history as more commonly and traditionally understood.

In short, pre-Adamite theorists view the biblical record as literal from Genesis 1:2 on. They see a large gap of time between Genesis 1:1 and 1:2. The theory, they believe, enables them to maintain the "veracity" of Scripture while accounting for the observed geologic record, at least as that record is interpreted by secular scientists. This view was particularly popular in the late 1800s, as theologians were faced with increasing ridicule from the scientific community for clinging to outdated biblical beliefs despite increasing evidence from geology, archeology, astronomy and paleontology, all interpreted in ways that contradicted the biblical account of origins and early life on earth. Over time, the utter lack of biblical support for the whole pre-Adamite protology lead to it becoming progressively marginalized. Very few proponents remain. Those Christians who accept the secular interpretation of the fossil record have no real alternative than some form of theistic evolution.

The Day-Age Theory

One big challenge, however, for theistic evolutionists is reconciling the biblical texts with the eons of time requisite for every evolutionary model of origins. One attempt to solve this problem is called the "Day-Age Theory," which has become popular among some Christians.

"Day-agers" take what they believe is a very literal view of Scripture. However, their literal interpretation hinges on what one could argue is a very symbolic spin on the word "day," which appears in each of the six days of creation as depicted in Genesis 1. Rather than a 24-hour period, as "day" commonly means, they see it as signifying a vast time frame, an epoch or an eon that could encompass millions, or even billions, of years. On the surface, at least, it's not an outrageous way to look at the word "day" itself. After all, we commonly say things like, "Back in the *day* of the American Revolution" or "In the old *days*" and so forth. The question is, however, does this understanding work here? Is this idea what the Bible intended to convey?

In their thinking, this interpretation could encompass the idea of God, as Scripture says, supernaturally creating the world and life on it, but doing so gradually, over vast stretches of time (certainly not in six literal days). Such a scenario would allow for distant starlight to reach the earth, as well as help explain the development of the fossil record and long geologic time, again according to the interpretations of secular geology.

This view, along with the others above, is, however, another abdication to the scientific community's materialist-only interpretation of the natural world, with all its assumptions, premises, and, to quote Lewotin above, its "unsubstantiated just-so stories." Day-agers compromise with Scripture in order to make it fit with the prevailing interpretation of the fossil record, the version which teaches that it was set down over long eons through purely natural processes. There are, however, many difficulties with this idea, one being how plants, created on the third day, could have survived for millions of years before the arrival of the sun, depicted as coming on day four.

Hugh Ross,[26] with his ministry called Reasons to Believe, is the most prominent leader of this movement. Other names promoting this view, in one version or another, are Gleason Archer,[27] Norman Geisler,[28] and Bernard Ramm.[29] Ross, in particular, pushes the day-age idea, which greatly undermines the integrity of Scripture.

The Young Earth (and Universe) Creationists

Then there are the "Young Earth (and Universe) Creationists," usually called "Young Earth Creationists" (YEC), though that word "Young" here should include not only the earth but the universe, as well—an important addition. The Bible and the Bible alone is the final authority for those people, at least that is what they claim.

YEC promoters view Genesis 1:1—"In the beginning God created the heavens and the earth"—as an all-encompassing statement. They believe that the word "heavens" refers to the entire universe and the phrase, "in the beginning," refers to 6000 years ago. Then everything that follows this verse describes the creation process, *not just of the earth but of the entire universe itself!* Most YEC scientists contest the Big Bang on principle, and the assertion that the universe is 13.8 billion years old. This position puts them at odds with basic cosmology: the very solid understanding that light from distant stars has been traveling for millions, or (in some cases) billions, of years.

They argue, in fact, that the universe and earth are *only 6,000 years old*—and they point to the Bible as their "proof." They believe that the earth was created along with the rest of the universe six millennia ago. When Genesis 1:2 says "the earth was without form, and void; and darkness was on the face of the deep. And the Spirit of God was hovering over the face of the waters," it is, they argue, a poetic way of stating that God's presence was here at the immediate creation. It was not talking about a tohu va vohu "without form and void" empty rock in space covered with water that was already here before God began the six-day creation process and rested on the seventh, even though that is, exactly, what the text says.

God then proceeds (according to their cosmogony), on day one of the universe to establish light; this day is the first time that light appears in the *entire universe.* Until then, there was no universe. This position is highly problematic biblically because the text also says that God divided the light from the darkness, with the light called "day" and the darkness "night," suggesting that this is a local issue, not universal. (What exactly is "night" and "day" in the entire cosmos? How does one envision that?)

On the second day, God's attention returns to this earth alone—and He then creates the atmosphere above the seas. Perhaps this is the air and sky as we know it, though many theories exist about exactly what is happening here. On the third day He separated the water from the land, establishing the seas and continents as well as the plants that grow on the land. On the fourth day, He creates the sun and moon and all of the stars in the entire universe, which explains why YEC promoters reject the idea of starlight having travelled long distances over millions and billions of years. It just appears that way to us, that's all.

Many YEC believers also suggest that God created all of the angels in heaven on the fourth day. On the fifth day God created the birds and fish. He completed His work on the sixth day, with the creation of land animals and with mankind, who, alone, was made "in the image of God." Some YEC suggest that angels may have been created on this day rather than on the fourth. Then, finally, God rested a literal 24 hours on the seventh day (though, as Christians who ignore the seventh-day Sabbath must, most YEC downplay this crucial fact).

Though much about this view does appear to harmonize with the Bible, the inclusion of the entire cosmos in the six-day creation leads to insurmountable problems, not just scientifically or logically, but biblically as well. For instance, Job 38:4-7 reads:

Where were you when I laid the foundations of the earth?
    Tell Me, if you have understanding.
Who determined its measurements?
    Surely you know!
    Or who stretched the line upon it?
To what were its foundations fastened?
    Or who laid its cornerstone,
When the morning stars sang together,
    And all the sons of God shouted for joy?

The text makes it very clear that intelligent created beings already existed at the time of the earth's creation, which does not fit the idea that they were created during the six days. Also, are we to believe that the newly created Lucifer sinned and deceived a third of the angels in heaven during the creation week? Did Lucifer fall and then deceive so many angels on the day that he was created? And when "war broke out in heaven" (Revelation 12:7), it did so on the fourth day? Had God's created cosmos fallen into evil even before God finished creating it?

Apart from the 6,000-year-old universe, and the numerous problems it creates, the rest of YEC, including the fall of Adam and Eve, their expulsion from Eden, the global flood, and so forth—all pretty much follow Scripture.

The Young Earth Creationist movement has developed momentum over the past couple of decades.[30] Institutions such as the Institute for Creation Research and Answers in Genesis have pushed conservative Christians to view Scripture as their authority, which is good. But, unfortunately, the 6,000-year-old-universe position, unnecessarily and (frankly) unbiblically, hurts creationism in general, because many biblical creationists, though accepting a six-literal-day creation, are aghast at the idea of including in that time period the creation of the entire cosmos. (We will look more at this in Chapters 4 and 5.)

The Puzzle Revisited

Again, like my family's Thanksgiving jigsaw puzzle, we know what the outcome will look like. Our question is how we get there. What we are seeking to piece together is the truth about our created world, however damaged by sin it has been. Our goal is to put the pieces of the puzzle together, the pieces that work, that lead us to answers that fit to create the complete picture.

Our world, of course, is a lot more involved, a lot more complicated, than any jigsaw puzzle sitting atop a family table. And it is filled with infinitely more pieces. But they have come together, obviously (or we wouldn't be here to muse about it); our desire, by God's grace, is to understand how the evidence—biblical, scientific, empirical and rational—brings together the pieces into a unified picture of the universe, of earth, and of God.

We continue toward that goal.

CHAPTER 2

The Old Universe/Young Earth Creationism Theory

Of old You laid the foundation of the earth,
and the heavens are the work of Your hands.
— Psalms 102:25

Centuries ago, someone asked what many deem the most basic question of all, the foundational question out of which every other question has its root. The question? "Why is there something instead of nothing?" That query goes as far back as "far back" can go.

In the last chapter, we looked at a few attempts to answer that question, not just in terms of all that exists—that is, the origin of the universe itself—but of what is of more immediate concern, which is our own origins as human beings here on the earth. After all, how can we know the right way to live if we don't know how, or why, we got here to begin with, or what the purpose of our lives is?

Our origins hold the answers to these basic questions. This is important because getting origins wrong can mean getting a lot of other things wrong too, as this quote reveals: "There is really nothing very funny," wrote theologian Peter Berger, "about finding oneself stranded, alone, in a remote corner of a universe bereft of meaning—nor about the idea that this fate is the outcome of the mindless massacre that Darwin, rather euphemistically, called natural selection."[1]

And because the models presented in Chapter 1—Abiogenesis, Theistic Evolution, the Active Gap Theory, the Day-Age Theory, and Young Earth (and Universe) Theory—did not work, scientifically, biblically, or both, this chapter explores an alternative theory, one that works not only scientifically and biblically but that also happens to offer a lot more hope than what Berger expressed above. This theory retains a literal reading of the Bible; in other words, it takes what the texts say about creation (and everything else) not only seriously, but accepts that the Bible means what it says. This view is much more consistent with the Scriptures than are any of the other origins views looked at so far. We call this view the Old Universe/Young Earth Creationism Theory.

So, what is the Old Universe/Young Earth Creationism Theory, and what are the biblical references that buttress it?

The Creation of the Cosmos

Why is there something instead of nothing? Because the self-existing, eternal God created everything that was created.

Unlike us, contingent beings who conceivably could not have ever come into existence, God is necessary and self-existent. By nature, He must exist. It's illogical and irrational that He not exist. He is not simply one being among other beings. He is Being Itself, the necessary foundation of all created reality. What this also means is that, unlike the universe, which exists only because God made it, God's existence is not contingent on anything else. God simply is, because God simply must be. There is no other logical explanation for the creation.

Upon this premise, that of a self-existent Creator, the Old Universe/Young Earth Creationism Theory teaches that, perhaps billions of years ago (current science estimates the creation of the universe at about 13.8 billion years ago, but that number will likely change—again), God created the entire cosmos, created all that was created. As John says about Jesus, "In the beginning was the Word, and the Word was with God, and the Word was God. He was in the beginning with God. All things were made through Him, and without Him nothing was made that was made" (John 1:1-3). In other words, whatever "was made," that is, whatever once didn't exist but then came into existence ("was made"), did so through Jesus.

When the Bible opens with, "In the beginning God created the heavens and the earth" (Genesis 1:1), though that obviously includes our earth, it could also mean the entire cosmos itself, even billions of years ago. The next verse says: "The earth was without form, and void; and darkness was on the face of the deep. And the Spirit of God was hovering over the face of the waters" (Genesis 1:2). How long the earth remained tohu va vohu ("without form and void") the text does not say. The earth may have remained without form and void for billions of years. For this reason, some have coined this theory the "Passive Gap" theory. This suggests that a large gap of time occurred between Genesis 1:1 and 1:3, but that no active creation of life and evolution occurred during that gap as proposed by the Active Gap theorists. Only in the next verse, verse 3, does the specific chronological sequence of "And God said . . . and it was so" begin. This is the sequence of a six-day creation and God resting on the seventh.

Yes, God is our Creator (Isaiah 40:28; Romans 1:25), but not only our Creator but the Creator of all life in the cosmos. Even without all the biblical evidence for other intelligent life in the universe, the sheer size and scope of the cosmos has caused secular scientists, such as those at the SETI (Search for Extra-Terrestrial Intelligence) Institute, to assume not only the existence of life out there but to look for it as well. And not just for microbes or the like, but for intelligent beings who, ideally, can communicate with us.

Besides the estimated two trillion galaxies, all created by God, the Bible is filled with references to intelligent extra-terrestrial life, all created by the same God who made us; and these beings were, likely, created before we were as well. What follows are some Bible texts that clearly and unambiguously teach about life in other parts of the universe.

"For we do not wrestle against flesh and blood, but against principalities, against powers, against the rulers of the darkness of this age, against spiritual hosts of wickedness in the heavenly places" (Ephesians 6:12).

"For by Him all things were created that are in heaven and that are on earth, visible and invisible, whether thrones or dominions or principalities or powers. All things were created through Him and for Him" (Colossians 1:16).

Depicting a heavenly judgment, Daniel 7:10 reads: "A fiery stream issued and came forth from before Him. A thousand thousands ministered to Him; ten thousand times ten thousand stood before Him. The court was seated, and the books were opened."

"And suddenly there was with the angel a multitude of the heavenly host praising God and saying: 'Glory to God in the highest, and on earth peace, goodwill toward men!'" (Luke 2:13-14).

There are, indeed, hundreds of texts that refer to angels, clearly beings who did not originate here but came from another part of the creation. The first two chapters of the book of Job show, in graphic terms, not only the reality of intelligent cosmic beings, angels, but their definite interaction here on earth. From Genesis to Revelation, angels are found in the Bible and, at times, playing key roles too.

Rebellion in the Cosmos

Not only are these beings depicted as interacting here, some also are clearly malevolent. That is, they are hostile to us. One of the earliest books of the Bible, Job, reveals Satan inflicting great suffering on Job. The last book of the Bible, Revelation, is explicit as to how broad that interaction is: "And war broke out in heaven: Michael and his angels fought with the dragon; and the dragon and his angels fought, but they did not prevail, nor was a place found for them in heaven any longer. So, the great dragon was cast out, that serpent of old, called the Devil and Satan, who deceives the whole world; he was cast to the earth, and his angels were cast out with him" (Revelation 12: 7-9).

War in heaven, evil on earth—how could this happen in a universe created by a loving God? The answer is the same: the freedom inherent in love. God is a God of love. Just as love exists in the Godhead, between the Father, the Son and Holy Spirit, love is a grand moral metaphysic of God's entire creation, and for intelligent beings—either scattered around the cosmos, or scattered on the surface of the earth—to love God they must be free not to love Him. Love, to be love, must be freely given, or it is not love. God can force all the cosmos to worship Him but cannot force a single intelligent being to love Him.

Against this backdrop, one of the greatest of these created beings in another part of the cosmos, known as Lucifer (Isaiah 14:12), for some reason rebelled against God. One text that helps explain is Ezekiel 28:15, which talks about his rebellion in general. "You were perfect in your ways from the day you were created, till iniquity was found in you." First, the Hebrew verb "created" (from *bara*) comes from the same word in Genesis 1:1 when God "created" (from *bara*) the heavens and the earth, a verb that appears *only with God as the subject.* Only God. Second, how could a "perfect" being created by a perfect God in a perfect heaven have iniquity? Again, the freedom inherent in love demands not the act of rebellion, but only the potential for it.

Over time, perhaps millions of years (we don't know), Lucifer exercised this free will and developed selfishness, pride, and covetousness. Another hint is found in Isaiah 14:13-14 which, talking about Lucifer, reads:

For you have said in your heart:

> "I will ascend into heaven,
> I will exalt my throne above the stars of God;
> I will also sit on the mount of the congregation
> On the farthest sides of the north;
> I will ascend above the heights of the clouds,
> I will be like the Most High."

As we already saw in Revelation 12:7-9, this rebellion lead to what the Bible called "a war in heaven" that, obviously, had reached down to earth as well, which helps explain, for example, what Paul wrote here: "For we do not wrestle against flesh and blood, but against principalities, against powers, against the rulers of the darkness of this age, against spiritual hosts of wickedness in the heavenly places" (Ephesians 6:12).

God, of course, could have wiped out Satan and "his angels" right from the start, but, again, motivated by love, He will allow the fruits of rebellion and evil to play out to their climax. When it's all over, and when all intelligent beings in the cosmos have seen the justice, love, and goodness of God—what happens? Then "at the name of Jesus every knee should bow, of those in heaven, and of those on earth, and of those under the earth, and that every tongue should confess that Jesus Christ is Lord, to the glory of God the Father" (Philippians 2:10-11).

The Creation of Life On Earth

At some point in the past, either before the rebellion or after (we have never been told), God the Father, the Son and the Holy Spirit decided to create intelligent beings, human beings, in His image, and do so on a rock of granite and water, the earth, floating in space in an arm of a rather large (but not the largest) galaxy, what we now call the Milky Way. Again, how far back in time God created this *tohu va vohu* rock, whether billions of years ago or just 6,000, the Bible does not tell us. (We will explore this issue more in the later chapters).

What we do know is that, however far back in time Genesis 1:1-2 was covering, Genesis 1:3 begins our only official account of the literal six-day process that God used to create the world and all life on it. However cryptic the texts themselves, however full of mysteries, and however much we don't know—enough has been given to show us that creation was the deliberate act of a loving God, with no room left for chance or happenstance, contrary to every popular evolutionary model, secular or theistic.

On Day 1, God's presence ("God is light") established electromagnetic radiation in this portion of the universe. Light, then, was established on Day 1 of creation on this planet, our earth. If God created the cosmos itself billions of years ago, then obviously light existed way back then as well. Here, though, in the immediate context of earth, there was now night and day, light and darkness. "God called the light Day, and the darkness He called Night. So, the evening and the morning were the first day" (Genesis 1:5). A "day" is an evening and a morning, or a literal 24-hour-time period.

Of course, there's the obvious objection, "How could there be night and day before the appearance of the sun?" A day does not require the presence of the sun. It requires only a revolution of the earth. If God's presence was hovering over the face of the earth and God is light, a revolution of the earth would have occurred during the first day and all parts of the earth would have been exposed to the presence of God and thus light, and all parts of the earth would have been shadowed from God's presence and light and would have experienced "night."

On Day 2, God rearranged the water distribution on the planet. "Then God said, 'Let there be a firmament in the midst of the waters, and let it divide the waters from the waters.' Thus, God made the firmament, and divided the waters which were under the firmament from the waters which were above the firmament; and it was so.

And God called the firmament Heaven. So, the evening and the morning were the second day" (Genesis 1:6-8).

According to the "Hydroplate Theory" (see next chapter), He did this by placing half of the water about 60 miles below the surface of the earth and left the other half on earth's surface. He created the atmosphere above the earth as the sky, "the firmament," and placed oxygen and nitrogen there to support the air-breathing life that He knew was soon to follow. Again, according to this theory, the water 60 miles under the earth served as a giant underground ocean that shifted with the lunar orbit, creating subterranean tides. These tidal flows created the bending of the earth's surface, and the heat generated produced warmth to the land resulting in evaporation of water from earth's surface every evening, which allowed for a watering of the vegetation without the need for rain. "For the LORD God had not caused it to rain on the earth, and there was no man to till the ground; but a mist went up from the earth and watered the whole face of the ground" (Genesis 2:5-6).

On Day 3, God separated the land from the seas. "Then God said, 'Let the waters under the heavens be gathered together into one place, and let the dry land appear'; and it was so. And God called the dry land Earth, and the gathering together of the waters He called Seas. And God saw that it was good" (Genesis 1:9-10). Though we don't have a lot of details, He probably established a single worldwide continent with many "seas" scattered throughout earth's surface. All of the water on the earth on the surface, and at 60 miles of depth, was (according to the theory) fresh water at this point.

Meanwhile, on this same day, He created all the vegetation, plants, trees, grass—everything like that. "Then God said, 'Let the earth bring forth grass, the herb that yields seed, and the fruit tree that yields fruit according to its kind, whose seed is in itself, on the earth'; and it was so. And the earth brought forth grass, the herb that yields seed according to its kind, and the tree that yields fruit, whose seed is in itself according to its kind. And God saw that it was good" (Genesis 1:11-12).

Day 4 is depicted like this: "Then God made two great lights: the greater light to rule the day, and the lesser light to rule the night. He made the stars also. God set them in the firmament of the heavens to give light on the earth, and to rule over the day and over the night, and to divide the light from the darkness. And God saw that it was good" (Genesis 1:16-18).

Though a lot remains unexplained, a few points seem reasonable from our perspective today. First, God put the sun and moon into relationship with the earth, giving the earth a perfect 360-day orbit around the sun, and the moon a perfect 28-day orbit around the earth (which have been altered since). The planets were likely formed now too. The stars outside of our solar system were most likely part of the original creation, perhaps billions of years ago. (In Chapter 4 we will discuss the biblical support for this idea further, as well as look into the theological implications of God's eternal being and creativity). Thus, when Genesis says, "He created the stars also," this means that He created the stars as well as the sun and moon, but at a different time (see Chapter 4). Because the stars are not especially relevant to the history of the earth, the details of their creation are not mentioned in the account—other than in Genesis 1:1, when "God created the heavens and the earth."

Day 5 continues with God creating fish and birds, all in order to fill the sea and sky that He created on Day 2. It reads like this: "Then God said, 'Let the waters abound with an abundance of living creatures, and let birds fly above the earth across the face of the firmament of the heavens.' So God created great sea creatures and every living thing that moves, with which the waters abounded, according to their kind, and every winged bird according to its kind. And God saw that it was good. And God blessed them, saying, 'Be fruitful and multiply, and fill the waters in the seas, and let birds multiply on the earth'" (Genesis 1:20-22). Notice, not a hint of evolution, not a hint of a gradual process, and not a hint of long ages—all indisputable details showing that evolution, including (even especially) theistic, cannot in any way be harmonized with any honest reading of the text.

Day 6 was the day that God created the animals and mankind.

> "Then God said, 'Let the earth bring forth the living creature according to its kind: cattle and creeping thing and beast of the earth, each according to its kind'; and it was so. And God made the beast of the earth according to its kind, cattle according to its kind, and everything that creeps on the earth according to its kind. And God saw that it was good. Then God said, 'Let Us make man in Our image, according to Our likeness; let them have dominion over the fish of the sea, over the birds of the air, and over the cattle, over all the earth and over every creeping thing that creeps on the earth.' So God created man in His own image; in the image of God, He created him; male and female He created them" (Genesis 1:24-27).

The angels in heaven, and all of the other "Sons of God," as described in the first chapter of Job, were previously created at a separate time and place not discussed in detail but alluded to elsewhere in the Bible (see Chapter 4). Meanwhile, what is depicted is the specific creation of humanity itself. God created man from the dust of the ground on Day 6, and then created the woman from Adam's rib on that same day (see Genesis 2). Man was formed in the image of God. He was created with the life-giving spirit, the "breath of life," which God Himself instilled into the lifeless body of man. "And the Lord

God formed man of the dust of the ground, and breathed into his nostrils the breath of life; and man became a living being" (Genesis 2:7). This, too, was in utter contrast to any evolutionary theory—theistic, atheistic, whatever—which would never conceive of, or imagine, a fully formed human existing before becoming alive.

Day 7 was the day, according to the Bible, that God rested from all of His labors.

"And on the seventh day God ended His work which He had done, and He rested on the seventh day from all His work which He had done" (Genesis 2:2). What's unique about this concept of a seven-day week is that it is, basically, arbitrary. Unlike each day itself, based on an evening and morning, 24 hours; or unlike a month, tied roughly to one cycle of the moon, 29.5 days; or unlike a year, based on earth's orbit around the sun, 365.25 days—the seven-day weekly cycle is not linked to any astronomical cycle or phenomenon. It's something that comes directly from special revelation, the Bible, which right at the creation shows that God Himself created not only the seven-day cycle but sanctified the seventh day, the Sabbath, as a sign of His creative power, a truth reiterated in the Fourth Commandment. "For in six days the Lord made the heavens and the earth, the sea, and all that is in them, and rested the seventh day. Therefore, the LORD blessed the Sabbath day and hallowed it" (Exodus 20:11).

The Fall

Why, then, is there something instead of nothing? Because an eternally existing God created everything that has been created; that's why there is not just "something" but *everything*. And Genesis 1 and 2 represent the Lord's "official" depiction of that creation, particularly of the earth and of life on it. This is how we know why we are here.

What we do not know is how long Adam and Eve lived in the Garden of Eden before the rebellion in heaven spread to the earth, and did so directly because of their sin. As perfect beings fresh from their Creator, they would not have had a problem with pregnancy, and so the Bible's silence about them having children prior to the Fall might indicate that they had not been in Eden very long (however one wants to define "very long").

What happened next was clearly depicted in Genesis 3. Separated from her husband, Eve had ventured into the portion of the Garden containing the Tree of Knowledge of Good and Evil, which was part of the original creation. "And out of the ground the LORD God made every tree grow that is pleasant to the sight and good for food. The tree of life was also in the midst of the garden, and the tree of the knowledge of good and evil" (Genesis 2:9). Of all the trees they could eat from, only this was prohibited for food (Genesis 2:17). Though for millennia people have speculated about the purpose of the tree, it should be obvious that whatever else might be involved, it was a test of their loyalty as free beings, a loyalty that became especially important considering all the chaos in heaven.

As we know, too, deceived by the serpent, Eve ate from the forbidden tree, and then she handed the fruit to Adam, who ate it as well (see Genesis 3:1-6)—and the rest is, as they say, history, the sad history of six thousand years of sin, suffering, and death. Death because, cast out of Eden, they did not have access to Tree of Life and what it offered. "Then the LORD God said, 'Behold, the man has become like one of Us, to know good and evil. And now, lest he put out his hand and take also of the tree of life, and eat, and live forever'—therefore the Lord God sent him out of the Garden of Eden to till the ground from which he was taken. So He drove out the man; and He placed cherubim at the east of the Garden of Eden, and a flaming sword which turned every way, to guard the way to the tree of life" (Genesis 3:22-24).

With the entrance of sin into this world, the story of redemptions ensues, first hinted at in Genesis 3:15. In this context, Old Universe/Young Earth Creationism then progresses forward, seeking to be faithful to the biblical text. However, the scientific debate with Young Earth (and Universe) creationists continues. For instance, just as they appeal to miracles to "explain" the distant-light time-travel anomaly (a fact of nature that contradicts, powerfully, the idea of a six-thousand-year-old universe), they do the same with the flood. It was purely "miraculous" and "supernatural," with no natural laws involved, arguing (for instance) that God supernaturally, *de novo,* created the water for the flood; and then, supernaturally, He just vaporized it away. We would argue that certainly, though the flood was a supernatural event, the Lord still could have and likely did still use some natural laws and processes as well.

The Bible records such miracles as the parting of the Red Sea (Exodus 14:21-32) and Jesus changing water into wine

(John 2:1-11). How did these and others, including Christ's raising the dead (John 11:38-44), occur? We don't know, but God can override the laws of physics and chemistry and even biology, laws that He established. In some cases, however, He can use these physical laws in order to accomplish His will, whatever supernatural elements are also involved. If we can search out how God used His physical laws as He interacts with His creation, we are much more likely to find His fingerprints there than we would with some kind of "God of the gaps" fallacy, claiming that it all is only and always supernatural all the way down. If some day we find that He did it this way, then we must all humbly bow to His omnipotence. However, in the meantime, we will search for God's working within the natural laws that He established, such as with the "Hydroplate Theory." And to that we turn next.

CHAPTER 3

The Hydroplate Theory

He gathers the waters of the sea together as a heap;
He lays up the deep in storehouses.
— Psalms 33:7

Two miracles are of special importance in the Old Testament, and they occur in the context of a major event in the history of God's ancient people.

The first miracle climaxed one of the most dramatic spiritual/theological accounts in the Hebrew Bible—the showdown in the ninth century BC between the prophets of the pagan "god" Baal, and the prophet Elijah, who sought to stem the tide of idolatry in Israel and to turn the people back to the true God, Yahweh. It came to a head on Mount Carmel, when Elijah publicly made this challenge to the priests of Baal: "Therefore let them give us two bulls; and let them choose one bull for themselves, cut it in pieces, and lay it on the wood, but put no fire under it; and I will prepare the other bull, and lay it on the wood, but put no fire under it. Then you call on the name of your gods, and I will call on the name of the Lord; and the God who answers by fire, He is God" (1 Kings 18:23-24).

After hours of shouting, screaming, pleading, even self-mutilating, the prophets of Baal didn't succeed in getting Baal to "answer by fire." In contrast, Elijah—after saturating the offering and altar with water—then cried out to his God. "'Hear me, O LORD, hear me, that this people may know that You are the LORD God, and that You have turned their hearts back to You again.' Then the fire of the LORD fell and consumed the burnt sacrifice, and the wood and the stones and the dust, and it licked up the water that was in the trench. Now when all the people saw it, they fell on their faces; and they said, 'The LORD, He is God! The LORD, He is God!'" (1 Kings 18:37-30). That was the first miracle.

The second miracle occurred shortly afterwards, and it is still part of the Elijah narrative; and though not as dramatic, it was even more consequential. Israel had been decimated by drought for three years. No rain had fallen, and the nation was in dire straits. Elijah then told his servant to gaze toward the west, to the sea; each time, however, nothing happened. But the seventh time, the servant came back saying, "There is a cloud, as small as a man's hand, rising out of the sea!" (1 Kings 18:44). What followed was the sky turning dark—and winds, clouds, and a heavy rain appeared, as Elijah knew they would. This was the second miracle.

How did both these miracles occur?

The first could have happened by God supernaturally forming an electrical storm overhead. We see this commonly where the weather patterns of an approaching storm are preceded by lightning and thunder. God may have used His omnipotent power to cause the elements of His creation to act at just such a particular time, and in just such a particular location. In short, God could have arranged it so that a powerful bolt of lightning consumed the altar and sacrifice on Mount Carmel.

He could have. However, that explanation does seem woefully inadequate, given what the texts say, and so most people who read this biblical story don't think of the miracle that way. This "fire from the Lord" certainly appears to be a totally supernatural phenomena, kind of a divine blow torch that created massive heat and energy, all in immediate response to Elijah's prayer that the true God reveal Himself in contrast to the false one, Baal.

Besides, there is no mention of a thunderstorm. The sky is clear (they had been in a drought). Lightning and thunder are mentioned elsewhere in the Bible, but are decidedly absent in this narrative, especially the sound of thunder, which would have definitely accompanied a lightning bolt of that capacity (consuming dust and stone?). The immediacy of the fire in response to Elijah's request, and the totality of the consumption, point to a purely supernatural act of God.

What about the second miracle, the "heavy rain" (1 Kings 18:45)? This could have occurred in the same manner as did the miracle of the fire coming down from the sky. The Lord, who could speak this world into existence, could have, and from absolutely nothing too, produced in the sky a dark cloud with a heavy content of water. He could have made it suddenly appear about 25 miles out over the Mediterranean Sea, and then supernaturally caused a strong wind to bring the heavy-laden cloud to the land and saturate a dried out and parched earth.

Or He could have produced the same effect—the cloud the size of a man's hand followed by the "heavy rain"—but in a different manner entirely. Months before the showdown at Carmel, the Lord could have changed the air pressure thousands of miles away over the Atlantic Ocean, which resulted in rain clouds forming and then, a few days prior to the upcoming drama, He moved those clouds into the Mediterranean in the direction of Carmel by means of the natural air currents. Then, minutes after the supernatural manifestation of fire from the sky, the small cloud appeared as Elijah's servant gazed westwardly for the seventh time.

Logic for this second miracle suggests (not proves) the latter mechanism as the most likely of the two. Why? Because if God wanted to suddenly pour water on the earth, He would not need a rain cloud to form over the ocean and make its way to land. Instead, He could have just supernaturally created water out of nothing and let it pour down, just as He (likely) created fire in the sky out of nothing and consumed the sacrifice and altar. Instead, He used natural means to end the drought. "Elijah was a man with a nature like ours, and he prayed earnestly that it would not rain; and it did not rain on the land for three years and six months. And he prayed again, and the heavens gave rain, and the earth produced its fruit" (James 5:17-18).

Whether the first miracle, or the second—we don't know what mechanism God used for these, or for any other miracles, actually. How did Jesus raise Lazarus's literally stinking corpse to life (John 11)? Care to make a guess? Or how did Jesus feed the five thousand with a few loaves and fish (Matthew 14)? We don't know. However, we can look at clues, as we did with the "heavy rain," that may provide tentative answers—in some cases, anyway. Context—such as events that happen immediately prior to and immediately after the miracle—may suggest a mechanism. After all, because God created the physical laws of nature, why would He not use those physical laws of nature to enact His will?

Flood Theories

With this background in mind, that is, the idea that God can use natural as well as supernatural means to accomplish His will—we will examine common theories about cosmology, creation, and the flood. To begin, we need to first look carefully at the biblical text that is talking about whatever event we are studying, and also at the relevant laws of nature, in order to determine, to the degree possible, what mechanism(s) was involved. In other words, might it not be more rational, and logical, to see if God used natural means to accomplish His ends, as opposed to conjuring up, by default, purely ad hoc supernatural explanations that overtly violate known laws of chemistry and physics? Is that certainly not worth a try, anyway?

For example, some theories of the creation account have come and gone like fashion trends, theories that often defy known laws of chemistry and physics. For example, popular in the creationist movement for years was the "Canopy Theory."[1] This theory suggested a canopy of water vapor surrounded the earth prior to the flood. This water vapor increased the air pressure and allowed plants and animals to grow. It also provided a more uniform temperature around the globe, which led (supposedly) to the plant and animal fossils that we see today in the arctic regions, meaning that in the past these places were not the frozen wastelands that they are now. According to the theory, at the time of the flood as depicted in Genesis, this water canopy collapsed, resulting in the deluge that destroyed the surface of the whole globe.

However, there were several problems with this "Canopy Theory" that eventually led to its demise.[2] One—the atmosphere cannot hold enough water to account for a global flood; that is, it is improbable, if not utterly impossible, for that much water to have been hovering in the air over the planet. Two—a water vapor canopy would trap heat to such an extent that the temperature on planet earth would rise and, essentially, cook every plant or animal on earth's surface. Life could not have survived in such an environment. Mathematical and engineering models have shown this fact to be the case. These and other objections to the theory have caused it to be largely abandoned by creationists. Among many other issues, the theory showed a lack of coherence with the laws of physics and heat transfer. That doesn't automatically disprove the theory, but this lack of coherence was one of a number of reasons that it has been set aside.

Two competing theories of the earth's creation and the global flood then replaced this canopy theory: Catastrophic Plate Tectonics[3] (CPT) and the Hydroplate Theory[4] (HPT).

CPT was developed by John Baumgardner, PhD, a creation scientist who worked for the Institute for Creation Research. He first proposed his theory in 1986 but then expanded it and applied a complex computer modeling in 2003. He suggested that the earth is composed of several crustal plates that sit on a bed of molten lava overlying the mantle. This lava moves, in convection currents, between the crust and the mantle. At the time of the flood, an unknown mechanism (it has been hypothesized, an asteroid coming into earth's orbit) caused a process of runaway subduction—when one tectonic plate sinks beneath another and descends into the earth's mantle—to occur. This subduction caused earth's plates to move at a pace of many meters per second, displacing the oceans, causing tidal waves, the heating of the oceans, and bringing the prolific rainfall depicted in the biblical flood narrative. The continental plates then crashed into one another, causing mountain formation. Sea floor spreading and volcanic activity on the ocean floor resulted, the remains of which we can see today.

CPT is consistent with the conventional theory of mountain and seafloor formation, which even secular scientists propose. The difference is in the amount of time required for these events to occur. CPT suggests the plates moved extremely rapidly, over a matter of hours or days. In stark contrast, the scientists behind common plate tectonics suggest, of course, that these plates moved slowly over millions of years. For example, the plates subducted over one another near the Pacific coast, raising the Sierra Nevada mountains inch by inch every year over vast periods of time,

Now, just because a theory challenges modern secular science is hardly a reason to reject it (especially when so much of modern science is based on a host of unproven and, often, dubious assumptions anyway). Yet CPT comes with numerous other problems, including one incredibly big one: its advocates appeal to *seven miracles by God* to sustain it. Seven![5] These include an asteroid the size of the moon coming into earth's orbit for two revolutions, and then leaving earth's orbit and returning to space. All this in order to create the mechanism that caused the rapid plate motion. A giant asteroid the size of the moon orbiting the earth twice, though possible, certainly seems, even to those who believe in miracles, as outlandish, if not utterly preposterous, speculation.

Two other supposed miracles, certainly not as farfetched as this first one, but both blatantly contradictory to laws of physics and chemistry, are: 1) God miraculously dissipated the heat generated from the plate friction and volcanic activity in order to keep the aquatic plants and animals alive. 2) Using thick molten lava as a lubricant, God miraculously reduced the friction caused by solid granite sliding for thousands of miles on another plate of granite. The lava, like a vast ocean of WD-40 household lubricant, allowed for the massive granite slides needed for the theory to work.

Could the good Lord have done these miracles? Of course, just as He could have miraculously created—out of nothing (and in an instant too)—the rain cloud, so crucial to the Elijah incident. But do we really need to resort to such speculations, which often make creationism lose credibility among those who, otherwise, would be open to a biblical account of origins?

The Hydroplate Theory (HPT)

In the remainder of this chapter, we will consider the Hydroplate Theory (HPT)[6] developed by Walter T. Brown, PhD, a speaker for the Institute for Creation Research (ICR) in the 1970s. He developed his theory in response to scientific problems that dogged other creationist models. Conflict arose within the ICR over which theory of creation and the flood should be presented at their meetings, leading to a split, not uncommon in the scientific community in many fields, creationist or secular. The rift over creation models continues to this day. Thus, we see very little reference to the Hydroplate Theory among creationist scientists from ICR and affiliated organizations, such as Answers in Genesis.

HPT, unlike some of the other models, relies on gravity alone to produce most of its effects; that is, it works within the known laws of physics, mostly. The HPT may require the intervention of God to initiate a crack in the surface of the earth, which began the cascade of events. However, once the process theorized by this model commenced, the laws of physics (all established by God) were set in motion, and the consequences of the flood played out upon the earth. The comprehensive nature of the theory to explain what we see on the earth is compelling.

Our theory will add some nuances to Dr. Brown's model. Dr. Brown is a Young Earth (and Universe) creationist and his Hydroplate Theory outlined in his book, *In the Beginning: Compelling Evidence for Creation and the Flood,* is written from this perspective. We will adapt his model to accommodate the Old Universe view discussed in the previous chapter.

Day 1. Our understanding of the HPT begins in Genesis 1:2: "The earth was without form and void, and darkness was on the face of the deep. And the Spirit of God was hovering over the face of the waters." Before God began His creative activity on this planet, about 6,000 years ago, there was no light here, but only a solid granite surface covered with a layer of fresh water. This water-covered granite planet was placed here on the edge of the Milky Way galaxy some indeterminate time in the ancient past by the Creator God as described in the previous chapter. Then God established light and energy, which were not here before His presence was made manifest on the earth.

Day 2. God separated the waters from the waters. Traditionally, this has been understood to mean that He separated water on the surface from water in the atmosphere. Hydroplate theorists suggest that this might, instead, mean that He separated the water on earth's surface from deep subterranean water. Half of the water was in a layer one mile thick, about 60 miles below the surface of the earth; the other half of the water was fresh water on the surface, still covering all the land. Recent discoveries have found more water under earth's surface than is located on its surface. It is trapped in a mineral compound called ringwoodite, a high-pressure variant of olivine. This may represent the residual water yet under earth's crust, left over after the expulsion of the liquid water at the time of the flood.

Day 3. He lifted the land out of the water. Picture the earth's crust like a 60-mile- thick ribbon with its deepest portions sitting on the solid mantle below. The highest portions sat above the surface water as dry land. Portions of the water would have been trapped between the peaks of dry land and formed inland seas. Other portions of the water would have been trapped below the surface of the crust as subterranean chambers–storehouses, as Scripture calls them. The crustal connections to the underlying mantle would be termed "pillars" supporting the earth's crust.

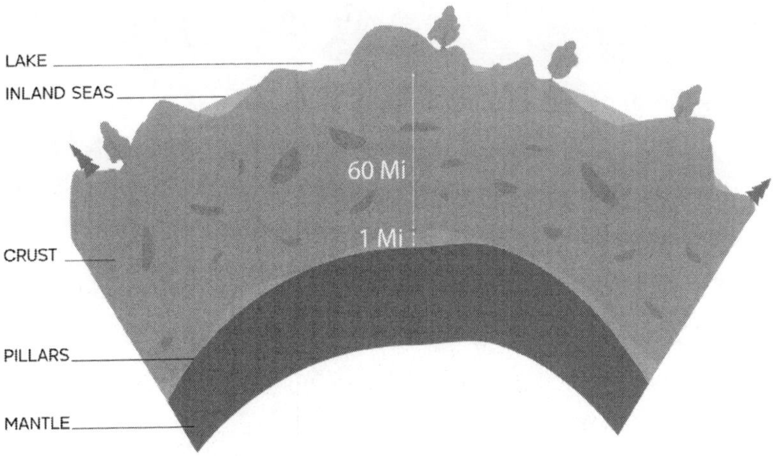

Day 4. God placed the sun and moon in their positions to warm the earth and to provide important stabilizing forces for His creation. God established a perfect 360-day rotation of the earth around the sun and He set the moon in a perfect 30-day

rotation around the earth. These perfect cycles would later be changed to a 365-day rotation of the earth around the sun and the moon to a 28-day rotation around the earth. The effects of the global flood would create these changes. The planets were formed, but there was no space debris such as asteroids or comets that could potentially harm God's faultless creation. Note, the stars outside of our sun were not created on this Day. According to our model they were created at a separate creation as we will discuss in the following chapter.

Day 5-7. The rest of the days dealt with the remainder of God's creative activity on the earth, culminating with Him resting on the seventh day, having "blessed" and "sanctified" it (Genesis 2:3).

Moving forward, Genesis 2:10-14 describes four rivers flowing out of Eden. "Now a river went out of Eden to water the garden, and from there it parted and became four riverheads" (Genesis 2:10). These rivers were, most likely, to provide water for those inhabiting the earth. If the earth were close to the same diameter as today, the circumference would have been about 24,000 miles. With each of the rivers flowing in divergent directions from Eden itself, they would have to flow for 12,000 miles to reach the polar opposite side of the planet, which was opposite the Garden of Eden.

Assuming some continuity with today, we know that the Mississippi River descends at a rate of half a foot per mile. Thus, a river in the Garden of Eden would need to begin at an altitude of 6,000 feet above sea level in order to reach the other side of the earth, at sea level. The Bible refers to the earth being founded upon the seas. "For He has founded it upon the seas, and established it upon the waters" (Psalm 24:2). It is likely that these rivers were connected by a series of tributaries and seas scattered throughout the planet. The Bible also says that there was no rain before the flood, but a mist rose each morning to water the earth.

Hydroplate Theory offers a plausible explanation for this mist. The moon in its orbit around the earth provided tidal forces (gravity) on the water here. Its gravity impacted not only the surface water, but also the water 60 miles deep in the mile-thick subterranean chamber. The twice-daily water motion caused deformation of the earth and created friction and heat on the granite crust, much in the same way that twisting a paper clip heats up the metal. This heat from the deep chambers of the earth would have been transferred to the surface, heating the land mass and causing a daily evaporation of the surface water, thus watering plants as the water condensed each morning. This cycle of water motion and flexing of the granite crust would have reached a steady state as God provided a mechanism for the earth to be watered, but also—and this is important: The cycle would open a plausible way for the flood to come should mankind—given the sacred (but risky) gift of freedom—fall into sin and wickedness, and God need to destroy the earth and most of what He had created.

Which, unfortunately, is precisely what happened.

The Flood

"Then the Lord saw that the wickedness of man was great in the earth, and that every intent of the thoughts of his heart was only evil continually. And the LORD was sorry that He had made man on the earth, and He was grieved in His heart. So, the LORD said, 'I will destroy man whom I have created from the face of the earth, both man and beast, creeping things and birds of the air, for I am sorry that I have made them'" (Genesis 6:5-7).

As evil prevailed among the antediluvians, the earth was filled with astonishing violence. Initially, the descendants of Seth maintained a connection to God and stayed faithful to the Lord. However, over time, the "Sons of God" (Seth's descendants, those faithful to the Lord) saw that the "daughters of men" (Cain's descendants, those who weren't faithful) were beautiful—and they took them as wives. This resulted in the righteous being contaminated with wickedness, and the number of God-fearing people dwindled to only the family of Noah and his sons. "Then the Lord said to Noah, 'Come into the ark, you and all your household, because I have seen that you are righteous before Me in this generation'" (Genesis 7:1).

What was the exact mechanism, or mechanisms, that God used besides rain? (Genesis 7:4, 12; 8:2). God may have simply, miraculously, caused a crack on the earth's surface, releasing the highly pressurized water below to burst forth from the subterranean chamber. Or (and this is admittedly, speculative)—after man sinned, the wicked descendants of Cain may have diverted the water flows for selfish purposes, and this may have led to portions of the earth's surface drying out over 1,500 years of time. The driest portion of earth's surface would have been the area for the first crack to occur. This view attributes the flood's initiation to being a consequence of man's wicked choices, as opposed to the direct judgment of God.

This is a similar question to the one posed at the beginning of the chapter, which dealt with how God ended the drought during the time of Elijah.

Regardless of what caused the initial crack in the earth, according to the Hydroplate Theory—the events occurring under the earth's crust were the same. "In the six hundredth year of Noah's life, in the second month, the seventeenth day of the month, on that day all the fountains of the great deep were broken up, and the windows of heaven were opened" (Genesis 7:11). The continued tidal forces of the water beneath earth's surface over time also increased subterranean heat. This heat on the water caused the water to increase in both pressure and temperature, and to take on a form called "supercritical water." This steam/liquid water combination is extremely erosive but also extremely lubricating. Rock, salt, carbon all dissolved, and fusion reactions developed radioactive heavy elements (a possible explanation for the radioactivity of some elements on earth).

The pressure and underlying erosive effects of the supercritical water, combined with the fracturing of the earth's crust from surface dryness and flexion of the crust, resulted in a massive fracture of earth's crust. The fountains of the deep, under immense pressure, burst—and poured out on the earth's surface.

The initial stages of the flood would have seen rocks launched from the subterranean chambers out of earth's atmosphere and into space accounting for the asteroids and comets we observe in our solar system today. This debris shot from the surface of the earth also accounts for the marks on the near side of the moon. The 60-mile-thick crust of the earth acted like a long gun barrel and accelerated the rocks to speeds necessary to escape earth's gravitational pull. The complete discussion of the physics needed for such items as escape velocities, heat transfer, and astronomical calculations of this space debris are beyond the scope of this book, but can be further explored in Walter Brown's book, *In the Beginning*. Further evidence for the plausibility of rocks being launched from the earth can be seen with the eruption of the Hunga Tonga volcano in 2022.[7] This underwater volcano erupted with such force as to launch water vapor 35 miles up into the stratosphere, halfway into space. The potential force of the Hydroplate model with the supercritical water, long-barrel (60 miles) passage of the rocks, and nuclear energy generated in the subterranean chambers all would have produced exponentially more force than the Hunga Tonga volcano, thus providing the necessary energy for escape velocities of earth material.

The water on earth's surface continued to rise after the rain subsided as water continued to escape from the subterranean space. Massive mud flows, with carbonaceous material, continually flowed over the earth's surface, coming from the eroded depth, and producing the multiple miles of layered sedimentary deposits seen on the earth today.

No other creationist mechanism can explain the miles of thick sedimentation on earth's surface. It is not possible for all of the sediment on earth's surface to have been held in solution and then deposited in multiple miles of thickness. The difference in mechanism can be conceptualized by envisioning a swimming pool with three feet of rock and mud at the bottom being filled by a large pipe of pure water flowing into it. This would have stirred up some of the dirt on the bottom of the pool and brought it into "solution," but—as time went by and the sediment settled to the bottom—a fairly thin layer of sedimentation would have developed. Contrast this with the picture of an empty swimming pool with a large pipe of mud flowing into it. The layering will be much thicker and much more layered with the mud flow than with the water flowing into the pool (as in the first model). One might anticipate more than half of the swimming pool would have been filled with layers of sedimentation after time had passed for separation of the water from the dirt. Earth's sedimentary layers are like a massive mud flow—but formed by new sediment *coming from under the earth's crust.*

With the water jetting out from beneath earth's crust, the 60-mile-thick plate would have undergone tremendous ero-

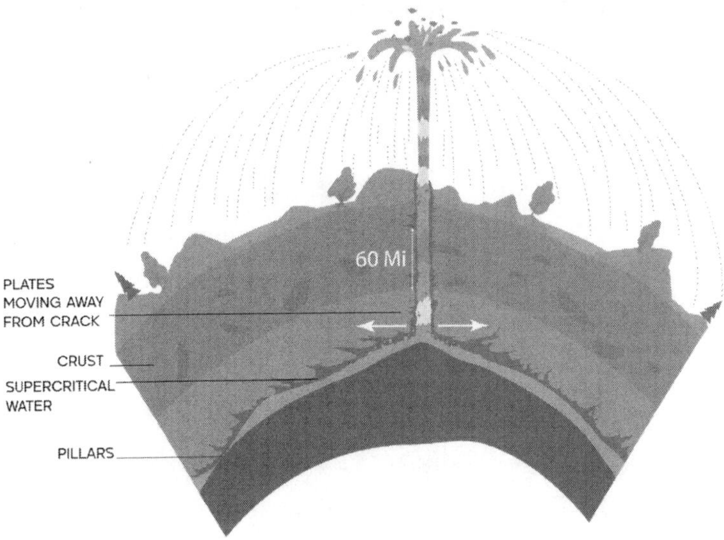

sion on the deep surface. The reduction in the mass overlying the earth's underlying mantle would have allowed the underlying mantle to bulge upward as the pressure above reduced. This mantle bulge, then forced the hydroplate to begin sliding away from the crack in the earth at speeds in the meters per second.

The supercritical water in the subterranean chambers created a lubricating effect allowing for the rapid motion of the hydroplate or earth's crust over the surface of the mantle. The plates then crashed into one another, lifting up mountain ranges in the same manner as when two cars collide and the hood of a car crumples from the impact.

Piezoelectric interactions—the physical phenomena where mechanical and electrical states interact within certain materials under tremendous pressure and temperature—resulted in fusion reactions that made heavy elements radioactive. (Prior to this, there would have been no radioactivity on the earth). This radioactivity was then scattered from the subterranean chambers onto the surface of the earth.

The water, now heavily contaminated with salt from the dissolving deep granite and also contaminated with deuterium (known as "heavy water"), spread onto the surface of the earth, creating the saline oceans that we now have. The water then flowed off of the uprisen land masses and created much of the typography of the earth's surface: mountains, valley, plateaus, and so forth.

The land masses then cooled. The oceans were warmer than they are today due to the impact of the superheated water coming from below, resulting in more evaporation and more precipitation on the land surface in the form of snow. This caused an ice age that lasted from the time of the flood until the days of Peleg (Genesis 10:25) —a few hundred years. Glaciation and features associated with it formed during this time period, as the receding glaciers carved out more valleys. Most of the ice sheets in the arctic regions today were laid down during this time period. During this glaciation, the ocean levels were likely 300 or more feet lower than today. This activity caused land bridges to form in the area of the Bering Strait, between Russia and Alaska, and in the archipelagoes of southeast Asia, allowing for more rapid migration of animals and peoples to all the continents.

Meanwhile, as the earth's diameter became smaller with the expulsion of a percentage of earth's mass, the length of a year changed to 365 days, from the 360 days at creation. Most ancient cultures hold to 360 days for a year, which is how long it takes for the earth to complete a rotation around the sun. The Hydroplate Theory postulates that the earth expelled from its surface a moderate amount of granite, water, and other material at the flood. This reduction in the mass of the earth would have compressed it to a smaller circumference and diameter, a reduction that has led to the earth spinning faster, much as ice skaters spin faster as they bring their arms into their bodies during a twirl. A faster spinning rate would have meant an increase in the number of spins the earth made in a year, increasing from 360 days (spins) at creation, to 365.25 days (spins) after the flood.

Also, some believe that the marked decrease in the long ages of humanity (they originally lived to almost a thousand years) resulted from these terrestrial changes. All water on the surface of the planet prior to the flood may have been pure water – H20. No neutrons in it. The post-flood oceans today contain water that has 150 parts per million of deuterium; that is, water contaminated with neutrons. Of note are the findings from samples of asteroids. The water found on these space rocks has two to three times this concentration of deuterium on it.[8] This is a much higher concentration than what we find in our current oceans.

It can be postulated according to the hydroplane theory that at the flood, when the waters of the deep were broken up, the deuterium-rich subterranean water was expelled onto earth's surface which was deuterium free. If the pre-flood subterranean water was in the range of 300 ppm deuterium, and the pre-flood oceans were deuterium free, and the water was half surface and half subterranean, then the resulting post-flood mixture would give us our current 150 ppm.

Research done by Laszlo Boros, PhD, retired now from University of California at Los Angeles, suggests that the deuterium concentration in our water may account for much of the aging occurring in our bodies.[9] Several other recent research articles are demonstrating similar findings in both clinical and animal models, both in vivo and intro studies.[10,11,12,13] This research is important to consider. Our body's cell membranes have mini motors that pump water in and out of them. When a deuterium molecule crosses those cell walls, it "clogs" up the machinery of the cells, causing damage that can have long-term effects. The ability of cells to repair damaged DNA and other cellular functions is inhibited, thus leading to disease and, subsequently, aging. This, along with other effects such as solar radiation and radioactivity together with the effects of deuterium contamination of earth's water supply, may have contributed to a reduced human life span of 120 years or less over time.

The Post-Flood Earth

Following the flood, the entire surface of planet earth was completely reformed and reshaped. There are no remaining rivers, buildings, cities, or other antediluvian remnants because the flood was a massive cataclysmic event that wiped them out, and every trace of them. Tens of thousands of feet of sedimentation was deposited on the surface of the earth, burying most (but not all) of pre-flood existence. For example, as the flood waters receded, they exposed things like dinosaur fossils that are found today, the remains of pre-flood animals.

Noah and his family landed in the ark on the mountains of Ararat (Genesis 8:4). One of his sons, Shem, spawned descendants who established the first post-Flood civilizations, which were southeast of Ararat in Mesopotamia. Some of Japheth's descendants moved up toward Europe, establishing the civilizations that we know as the Neanderthals and others. The descendants of Ham moved south, establishing the early Egyptian empires, building the pyramids, and settling on the Nile River.

The Hydroplate Theory, with the proposed additions, presents a coherent, explanatory outline for our planet, from inception to its present form. And most importantly (as we shall see), it is consistent with Scripture. That is, it does not rely on seven incredible miracles of God, or whatnot, in order to explain physical phenomena. Instead, it is comprehensive, covering many areas of known science. When combined with the Old Universe/Young Earth Cosmology, it gives us a logical overview of how God created our world, and what happened to it—and to us—after sin entered. The following is an abbreviated list of issues this theory addresses:

How did the hydrologic cycle work on the pre-flood earth?

Why wasn't there rain before the flood?

Why isn't a year exactly 360 days long and a month exactly 30 days long today?

Why are the oceans salty today?

Why did man live over 900 years before the flood?

How did the Ice Age occur?

How can there be tens of thousands of feet of sediment on the earth today?

Where did all the flood water come from and where did it go?

How did the topographic features of earth appear, including limestone and the Grand Canyon?

How did the continental plates, mid-oceanic ridge, and other seafloor features develop?

Why are there temperate-zone animals and plant fossils found in the arctic regions?

Where did radioactivity come from?

Why are fossils sorted as well as they are in the geologic column?

These and many other questions are answered with the Hydroplate Theory. Again, the Lord may have supernaturally caused a crack in the earth, but once this initiating event occurred, the subsequent events explained by the theory fall out as a result of the effects of gravity and known science. When comparing different scenarios of creation of the earth and the events of the flood, this theory offers a compelling explanation consistent with an Old Universe/Young Earth theory of origins.

In the next two chapters we will compare the theories we presented in Chapter 1 with the Old Universe/Young Earth/Hydroplate theories. For brevity, we will refer to our theory as Old Universe/Hydroplate or OUHP. The next chapter will look at the scientific issues, and the following chapter will look at biblical questions.

CHAPTER 4

Comparing the Scientific Evidence

For since the fathers fell asleep, all things continue as they were from the beginning of creation.
— 2 PETER 3:4

What follows immediately is a recap of the various theories of cosmogony and the development of the earth and life on it. We have looked at so far:

1) Secular Science (SS)

2) Theistic Evolution (TE)

3) Active Gap (AG)

4) Day-Age (DA)

5) Young Universe/Young Earth (YUYE)

6) Old Universe/Hydroplate (OUHP)

Many other theories beyond these exist in the area of earth science, and cosmogony in general, and many variations of each one do too. For example, most Hydroplate theorists—notably Walter Brown, who conceptualized the theory himself—believe that the earth and the universe were created from nothing *6,000 years ago*! He believes that the earth's mass was nonexistent prior to the creation account in Genesis 1. That view, however, remains extreme, nor is it necessary in order to be faithful to the biblical text. (In fact, it also happens to be unbiblical.)

In contrast, what we have done is take the Hydroplate Theory and applied it to our Old Universe model. This approach, we believe, is the best way to describe the scientific and biblical evidence for the creation, which includes the entire cosmos, as well as the earth and also the flood. Walter Brown's idea that the universe is young is dealt with in our analysis of the YUYE view. Thus, in this chapter we will look at these six theories and compare and contrast them with the available scientific evidence.

We will start with the scientific evidence for the theory of evolution.

Secular Science

Secular science (SS) argues that everything came from nothing. All matter in the universe arose from no matter. In other words, *everything comes from nothing*. But, for atheists, this is their only option. If the universe started from something, then whatever that "something" was would itself need an explanation, and on and on into an infinite regress. Their only out, other than nothing, is an eternally existing God. But because they a *priori* reject the idea of such a God, they choose "nothing" instead as the cause of the universe. Or they dance around the issue by stating things like, "Perhaps there is a cycle of universes, or multiple Big Bangs," which doesn't solve their problem because a cyclic universe, as do multiple Big Bangs, also needs a prior explanation as well. In short, atheists do not have a reasonable explanation for where they (and the cosmos) came from; they know only that they (and the cosmos) are.

SS stakes its reputation on taking the data and applying it to their theories. They have made great efforts to adjust their theories to new discoveries. Thus, the development of the Big Bang came about from the information obtained about the "red shift"—the stretching out of light waves observed in the cosmos. This flexibility of their theories has also caused difficulty for SS scientists. The James Webb telescope's recent discovery of fully developed galaxies 12 billion light years away is causing many to rework their theory of the Big Bang. If the universe began just 13.8 billion years ago as the current Big Bang scenario postulates, then it should be impossible for a fully developed galaxy to have formed with mature shape patterns and organization in only 1.8 billion years. The observation of light 12 billion light years away showing mature galaxies is an enigma. This has caused SS recently to look to new theories to explain the cosmos as they observe it from their paradigm.

Beyond just the difficulty of there being a lack of enough time to explain distant observed galaxies, an additional challenge for the atheistic evolutionists is that of spiral galaxies which have age characteristics that cannot be explained by the observable mass in the universe, at least not explained according to the present understanding of physics. Calculations have shown that a certain amount of mass is necessary in order to change the shapes of galaxies; the problem, however, is that there is not enough mass in the observable universe to account for the shapes of galaxies as they exist now, again at least according to the present understanding of physics. This is the reason that secular scientists have turned to the concepts of "dark energy" and "dark matter" (dark in the sense that we cannot see them), which they theorize had created the mass necessary for the shapes and distributions of galaxies as they appear now. In other words, because our present physics can't explain the phenomena that we see—rather than change the physics—the scientists assume that 95 percent of the universe is made up of this dark mass and dark energy.

Secular scientists claim that the earth developed from a condensation of various elements spread across the universe by supernova explosions scattering cosmic debris, in which lighter elements fused into heavier elements. These elements and molecules, including water, condensed into the sun, into this planet, and into the others in our solar system. Over time, these celestial bodies were brought into a gravitational rotation around our newly developed sun, which was really a massive collection of hydrogen atoms at the center of our solar system. All that is, at least, what the theory[1]—full of speculations and unproven, if not rather farfetched assumptions—teaches.

First, how could water condense on this planet and not vaporize into space due to the large amount of heat generated from the bombardment of asteroids and other high-energy objects hitting the earth? Why are there such varied compositions of the various planets in our solar system? The sun temperature and the earth temperature greatly vary, obviously, and, as discussed in the first chapter, this reality continues to be a challenge for this secular view. The distribution of elements on planet earth is an enigma, as well. Why are there heavy elements near the surface when the earth was previously a hot molten planet in which the heavy elements should have sunk to earth's core? These are just a few of the problems that secular science has no answers for, which means their theory is as much a leap of faith as are those who believe that God created our cosmos.

The real fun stuff begins with the existence of life on earth, itself a mystery that to this day defies the best of modern science (which still has not even come up with and agreed upon a definition of the word "life," much more, an explanation of how it started). Though theory after theory about the undirected and chance origins of life are proposed, each one crashes and burns until a new theory comes that, eventually, crashes and burns too. As the poet Campbell McGrath asked: "What breathes life into a handful of dirt and casts it staggering along the orbit of its fate?" Secular science does not know. Meanwhile, abiogenesis (the theory that life arose from non-life) goes further and further into crisis each year because, the deeper science probes life, the more it discovers that life is much more complex than previously believed. The irony should not be missed: The more science reveals about the nature of life, the more far-fetched become science's theories about how it started.

Chemical evolution itself, that process that supposedly led to life, had no driving force of natural selection to move it forward, either. That is, methane, ammonia, hydrogen, water vapor, some of the earliest chemicals that were (supposedly) to have led to life, were not "naturally selecting" for anything. They were just chemicals, and so the idea of natural selection applying to them is as ridiculous as saying that natural selection is what keeps a rock surviving. The lack of catalysts; the lack of preservation for developing chemicals; the austere environment of early earth—none of this was conducive to the chance formation of life. Scientists know this, too, but certain that their atheistic, abiogenic theory of a purely chance creation must be true, they keep on this false path anyway. After all, to even hint at design is blasphemy. All this—in addition to the mathematical impossibility of the development of life if one looks at random chance and the number of random reactions needed to move life forward—makes abiogenesis as likely an explanation for life's origins as is the Babylonian creation story.

The sheer implausibility of the whole atheistic abiogenic enterprise is further exposed by the challenges of what has been called "irreducible complexity." The archetypical example, first proposed by Michael Behe,[2] is that a mousetrap cannot be simplified from its five basic components. If any one of the five components were not there, that is, if even one were not part of the whole contraption, the device would never catch mice. It would not work at all. Without all the parts functioning, it would not be a mouse trap. Behe's point, and it's a good one, is that there is no natural way for the components to evolve without the presence of the other parts. How would it know what to evolve into? Issues like that only add to the incredulity many have toward atheistic evolution.

Another hurdle is sexual reproduction. If one can postulate that, with slow progressive changes, an organism can evolve over time, how does one postulate the existence of parallel organisms which develop separate sexual characteristics that, eventually, combine to propagate the species. How does this work again? A male evolves male genitalia, which includes semen, sperm, prostate, vas deferens, testes, urethra and so forth, while the female evolves her genitalia, which includes a womb, fallopian tubes, ovaries, eggs and so forth—and, though both are undirected, purely chance developments, they miraculously harmonize and come together to produce more beings, male and female, who continue the process. "Far-fetched" is too generous a term to depict it.

The geologic column is their ace in the hole, at least for secular science (or so it thinks). At the floor of the fossil record sits the Cambrian layer, filled with trilobites—sea creatures that live on the ocean bottom. They are considered simple ancient creatures. Only problem? Similar creatures exist today. The horse-shoe crab is anatomically very similar to the trilobite. We find these creatures living in the coastal water of North America and Asia today. The argument is made that animals such as trilobites have just persisted through the ages. If this argument is true, then we should see all animals found in multiple levels of the geologic column. The fact that we don't see marine floor animals in the upper levels of the geologic column suggests that the animals are located at their various levels because of sorting rather than progressive development over millions of years. Otherwise, we should find marine bottom-dwelling animals at higher levels of the column as we know marine-floor animals still exist today.

The complexity, size, and habitat for animals increases as the layers get nearer to the surface, buttressing the argument that they evolved into more complicated, more highly evolved creatures, which, of course, required long ages.

However, there are no locations with the entire geologic column present. In other words, it's a piece here and a piece there that they have to try and put together. The layers, in short, require quite a bit of interpretation and speculation. In fact, different theories come with their own interpretations of the evidence. Also, why are some animals that are found in the lower layers still present and alive today? That is, we have fish and we have cows. They both, obviously, exist today, but are found in different layers of the geologic column. So, is the column more an evidence of sorting or mobility than it is of an evolutionary progression in development? That certainly is a possibility. Animals may have been sorted according to their density, size, shape, habitat, ability to move away from danger, which, along with other factors, created the geologic column as we find it today.

The last area of scientific evidence we will look at in support of the secular scientific view is that of geomorphology—the things we find on earth's surface. Evolutionists postulate that mountains developed over time with the slow movement of tectonic plates pressing together to uplift land masses. Physics again works against this theory.

It is, however, difficult to conceptualize how these mountains, with fossils of marine animals on their surfaces, were pushed up from the seas millions of years ago. The problem is one of momentum. Visualize two parked cars facing one another, their front bumpers touching. If the drivers of the opposing vehicles both press the accelerators to the floor, we will witness burning rubber, but very little metal crumpling and front-end damage to the cars. We will see when we discuss the Hydroplate Theory that momentum is needed for mountain formation.

Meanwhile, erosion patterns on the earth's surface, and between geologic layers, are also features that are worth analysis. Secular science suggests that erosion such as the Grand Canyon and other valleys, and the large glaciation changes seen in northern latitudes, indicate long ages. It is apparent that some areas of the earth's surface erode more quickly than do others. If a surface is exposed to rain, ice, wind, animal migration, or other such interference, it is likely to deteriorate over time. Softer surfaces will erode more quickly than do hard ones as well. Running water will create channeling. Thus, we see hills and valleys, gorges and pits, all secondary to erosion. Again, Old Earth models that need millions of years predict that there should be very few sharp surfaces on the earth. Since mountain formation and surface erosion are occurring at the same time, we should see only smooth surfaces—which we don't. As with so many other geomorphic features, the sharp edges of canyons and pointed mountaintops point to a much shorter time span.

Another fantastic example of these erosion issues is seen in these photographs taken from a satellite of the Grand Canyon in comparison with the adjacent photograph of the beach taken after a wave receded from the Pacific Ocean. One can scarcely differentiate among the three photographs. The reason being, they were both carved by water a relatively short time ago.

The two photographs below are satellite pictures of the Grand Canyon.

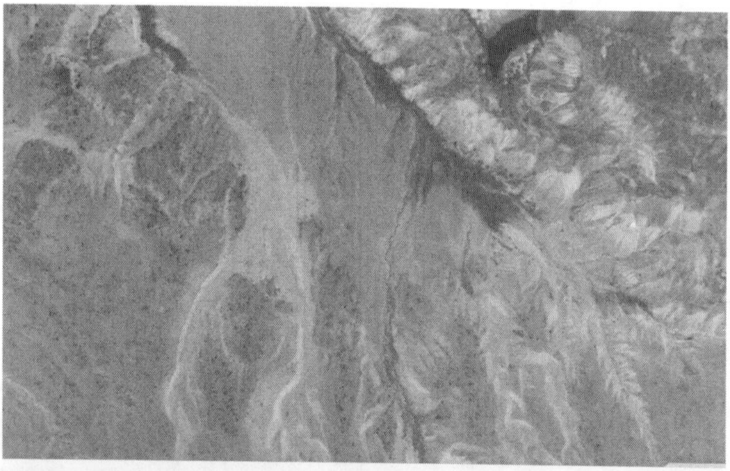

The picture below is taken from 4 feet above the beach in San Diego.

Additionally, in the Grand Canyon, the layers are sharply defined. If millions of years occurred between various layers, then we should see variability in the thickness of the layers, with evidence of erosion between them. We should see evidence of animal activity on a microscopic level, and also plant life having taken root between layers. Yet we see almost none of this, which bodes very ill for long age theories of any kind, in which all these life forms had plenty of time to come and go. Little erosion, and very little plant and animal activity appear for thousands of miles on layers in the Southwest United States, which should not be the case if millions of years are postulated to explain what we see. Erosion—both on earth's surface, and between the geologic layers—points toward a recent flood as the source of the layering, as well as remaining the best explanation of the current erosion patterns on the earth's surface.

Ice core analysis in Greenland and Antarctica suggests long ages of the earth and many tens of thousands of layers are seen in these ice core samples. The multiple layers observed, the identification of specific layers of volcanic ash associated with specific eruptions, the presence of CO2 gas concentrations in different levels consistent with the predicted CO2 present at differing times historically all support the long ages theory. Critics holding a young earth perspective counter that the layers may represent multiple snowstorms with many layers being laid down per year. They point to evidence that CO2 gas diffuses between layers and that the volcanic ash cannot be accurately corroborated with historical volcanic eruptions. Ice core analysis presently is one of the stronger arguments for secular science.

Theistic Evolution (TE)

In contrast to secular science are cosmogenic models that allow for an involved God in His ancient creation who, therefore, does not require dark energy or dark matter (which have never been seen) in order to explain what we can see: the shape and distribution of galaxies. These models do suggest that God supernaturally, over time, formed and shaped the galaxies. In other words, God is not bound by the need for dark matter or energy, or even by the limitations of aging matter. Whatever the great mysteries that remain about the formation and function of the cosmos (and there are many), God is not bound by the laws of nature that He has created; thus, He can, and does, supersede them when needed.

Theistic Evolution (TE) seeks to solve these problems by bringing God into the equation. He is the Source of all matter, of all the created cosmos (the only logical option for the creation, actually). Theistic Evolution, however, rapidly turns away from God as the author of all matter (which He is) to the secularist view (though some among them would deny it) for all subsequent actions in the universe after God first created the universe. They, as do the secularists, appeal to processes alone for the more complex development of matter, such as supernovas and their fission and fusion reactions, which they say account for the development of the more complex atoms beyond hydrogen, helium, and lithium. And, of course, with the exception of God having created some kind of early life form, many theistic evolutionists share the secularist view of evolution, with natural processes, such as random mutation and natural selection, accounting for the wide variety of life on earth, all of which arose from the supernaturally created early life forms, perhaps a single cell of some type or another? On that, atheist or theist, they can only guess. Thus, it is very difficult to evaluate Theistic Evolution from a scientific basis because they have no testable hypothesis by which to evaluate at which point God entered the creation equation. He may have entered at any point and solved any problem, or He may not have and the theory reverts to secular science.

Though not as ludicrous as its atheist cousin, Theistic Evolution continues down the same impossible pathway for the development of life, just throwing in the occasionally guiding hand of a theos (of one kind or another) who, for whatever reason, chose to create life through a long, violent, and torturous process that required billions of years of pain, suffering, and death. Possible, but this theory goes not only against any honest reading of Scripture, but depicts God as anything but benevolent and loving.

Active Gap (AG)

Turning now to the scientific evidence for the Active Gap Theory, this view relies heavily on a "gap" between the first and second verse of Genesis 1. Genesis 1:1 states, "In the beginning God created the heavens and the earth." Then long ages of time pass before God reappears in verse 2, just 6,000 years ago. The theory varies as to what occurred during those long ages between those verses. God may have taken the theistic evolutionist approach and stepped away to allow the process of evolution to develop the cosmos, earth, and life by a slow process. Or He may have de novo created the universe and the earth and life on earth and left it to eons of time of death, destruction, layering of geologic columns, and land formations until a point where He decided to return to place His Spirit in Adam just 6,000 years ago.

The Active Gap Theory suggests that, when God created the rock out in space (earth), He also placed on the earth plants and animals, perhaps even hominids. That is, an entire ecosystem and civilizations were present before God put His

Spirit in Adam and Eve, turning hominids into full-fledged human beings.

Though they point to the fossil record as evidence for their position, the puzzling thing is the lack of hominid fossils and lack of archeological evidence for hominid "societies," or for any kind of pre-human civilization. All this could have been destroyed in a global flood, yes; however, many Active Gap proponents do not believe that a global flood destroyed the earth, despite what the Bible teaches. Instead, they look to various local floods in order to account for the flood story of the Bible, despite there being nothing in the Genesis flood account depicting it as local, in contrast to every aspect of the biblical account depicting it as global. (One simple example: If the flood were only local, why build a boat? Why not just tell Noah et al. to move out of the flood zone?) Active Gap theories believe that millions of years of evolutionary processes were needed to account for the fossil layers, something that, like their denial of a global flood, goes against the Bible.

The AG Theory basically suggests that life was created by God twice. The first time, millions to billions of years ago; then, a second time, now at the creation depicted in Genesis 1-2, when He (supposedly) put His spirit in mankind, about 6,000 years ago. We have already stated the difficulties burdening this view, primary among them the lack of developing hominids in the fossil record. There is a paucity of ape-like animals in the development phase that was supposedly moving toward man. That is, though their fossil remains should be found everywhere, they are rarely found anywhere. We should see primitive tools (we don't), communities (we don't), and remnants of these early civilizations (we don't). It has been said that if one gathered all the hominid bones of these supposed pre-humans, they would all fit in the trunk of a car. What kind of car? It hardly matters.

They generally account for both the geologic column and geomorphology from the first creation, alluded to in Genesis 1:1. In this version of creation, billions of years ago God made plants, animals, and hominids as well as mountains, valleys, and oceans. And then, as those billions of years of death, destruction, and evolution had passed—all part of a loving God's creative process—the geologic column arose, though with the same interpretive challenges facing secularists and theistic evolutionists alike.

Day Age (DA)

The DA model relies heavily on the data obtained from secular science. This is the basis for why they suggest that the Genesis account must have been long ages of time. After all, the science demands it, they say. Thus, the universe with a very old appearance must be billions of years old. The science of distant starlight is consistent with this view. But the DA proponent believes in the authenticity of Scripture, and thus views God as active in all aspects of His creation. Thus, He was anciently active in the creation of the cosmos when on Day 1 He created light. But the galaxies we see were also created on this Day. This lasted for billions of years as the light and galaxy formation had to develop. This appears consistent with what our scientific observations are. Then on Day 2, He actively placed the earth in space with water on its surface. But the atmosphere had to develop and evolve over millions to billions of years. This was followed by Day (epoch) 3 during which the dry land emerged from the ocean and plants and animals were created (evolved). This theory now suffers from the same contradictions that secular science does: the challenges of temperature variability, of stability of life on the planet, of the unlikelihood of the entire evolutionary process, and the lack of consistent evidence in the geological column.

The DA theory places life on the earth on Day 3, after the atmosphere and oceans were established on Day 2. According to this theory, the life that was created was simply plant life. As already touched on, but worth fleshing out, if the following day was coming millions of years later, when the sun, moon, and stars were created, then how did all that plant life survive until then? They would have had to survive until the sun was firmly in the sky and burning just as it needed to in order to sustain them. In other words, plants would have had to survive for millions of years without sunlight (unless there was another miracle here as well in which God kept them alive without sunlight).

Plants, as we all know, need the sun. However, proponents of this theory, such as Hugh Ross, turn to modification of terminology in the Bible. He suggests that the sun was present, but just not visible; perhaps there was cloud cover. This too, however, would have inhibited plant growth. Billions of years of cloud cover probably would not have been conducive to the flourishing of plant life.

The birds and the fish were created on Day 5, and so would have been present for millions of years prior to the creation of land animals. The fossil record would have shown the birds to be located below the land animals in the geologic column. Yet that's not what the geologic column reveals, at least not consistently. Fish and birds should all be located well below the land animals, at least according to the Day-Age Theory. However, the fossil record shows a mixture, with many birds being located above land animals in the fossil record, a blatant contradiction to the theory.

The Day-Age Theory, though having some affinities with the Young Earth view, still comes with massive problems. There are Day-Age theorists who turn to Scripture as literal history—but only after Genesis 2. The motivation for this position is, basically, the same as secular science: viewing the geologic column as evidence for eons of time of death and destruction between the "day ages" of creation.

It is, frankly, a sad commentary that the DA theorists place the validity of secular science on a pedestal higher than that of biblical authority. The contortions necessary to marry the two opposing views of secular science and the Bible make the theory untenable. That is why the final two views we will evaluate place the authority of Scripture above that of the secular scientist and rather take science viewed through the lens of Scripture to establish their theories.

Young Universe/Young Earth (YUYE)

The Young Universe/Young Earth (YUYE) group struggles with the issue of the initial creation. They insist that God created the entire universe only 6,000 years ago. Thus, without His direct supernatural intervention in terms of light travel, they turn to untenable theories, such as White Hole Cosmology and Anisotropic Synchrony Convention.

White Hole Cosmology suggests that the universe was created in a "white hole" as opposed to a black hole. This is a theoretic, but unobserved, construct that proposes that time can be sped up almost infinitely, as opposed to it slowing down in a black hole. Therefore, if the universe were to have been created in such a cosmological situation, time would have been occurring outside of the earth incredibly fast while it was slow on the earth and therefore a day could have gone by on earth while the expanding surrounding universe was experiencing billions of years of time. Even Russell Humphreys, the theory's originator, has largely abandoned this theory.

Anisotropic Synchrony Convention postulates that light travels at different speeds depending on which direction it is traveling. Thus, the light that is traveling toward us is coming at us instantly, but light that is traveling away from us is traveling at half of 186,000 miles a second (the calculated round trip speed of light). This is based on the fact that the speed of light has only been measured in its round trip (there and back) passage. Of interest is recent work demonstrating the photographic passage of a laser beam in space using high-speed photography. Here is a photo from the YouTube video by Alpha Phoenix demonstrating this light passage.[3]

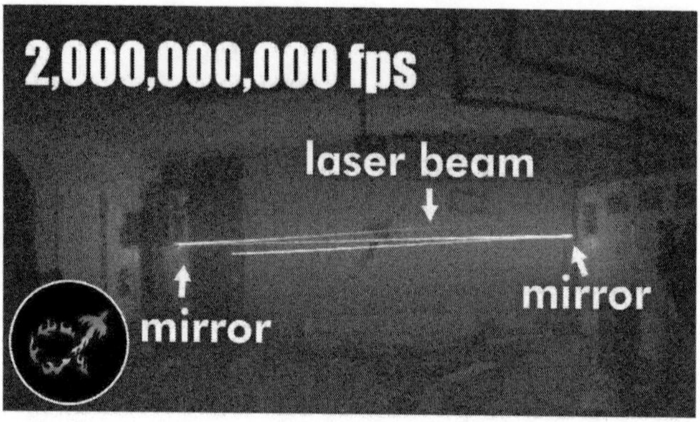

The camera, shot at 2 billion frames per second, clearly shows the passage of light over time in both directions. The presenter shows how moving the camera to the originating source of the light causes it to appear faster on the return trip, but the explanation is clearly presented demonstrating that the light is moving at a consistent speed in both directions and that it is only the camera's ability to capture it that changes its apparent speed. This presentation completely debunks the Anisotropic Synchrony Convention theory for the passage of distant starlight.

Both of these proposals, although interesting, stretch the reasoning mind beyond the level of plausibility. They are exceedingly far-fetched speculations that violate many established laws of physics—all in order to try and explain time/light issues in the universe. These theories are so difficult to apply that one of the leading astrophysicists of the YUYE movement, Danny Faulkner, was finally forced to argue that God supernaturally placed the light in space and, instantaneously and miraculously, advanced the light to be everywhere at once in the universe at the time of creation.[4] That's it, a God-of-the-gaps ploy that secular scientists disdain, and rightly so. Worse, these ad hoc arguments are the kind that make all creationists look bad, no matter how many creationists reject it.

The Young Universe/Young Earth (YUYE) theory teaches that the earth itself was created *de novo* 6,000 years ago. This view has posed a problem for more conventional dating methods, such as radioactive dating, which openly contradicts the YUYE theory. The RATE (Radioisotopes and the Age of the Earth) Project,[5] an an initiative of the Institute for Creation Research, the Creation Research Society, and Answers in Genesis, was established to deal with the question of radiometric dating. How accurate is it, really? The conclusion of this project was that accelerated decay rates were probably necessary to account for earth's radioactivity. This means that rapid radioactive decay is observed. Therefore, it must have been sped up by some process.

They also found that there may be heat problems related to this accelerated decay. That is, how did the earth get rid of the excess heat with the accelerated decay, which they propose occurred during the flood? This, and other questions, have arisen related to a young earth being created just 6,000 years ago.

In support of a recent creation of the earth are helium diffusion rates. Zircons, rocks found on earth, have a predictable rate of diffusion of the element helium. If these rocks are extremely old, millions to billions of years, then we should not see any helium in them. However, this is not the case. The amount of helium seen in zircons is consistent with a 6,000-year time period since their creation.[6]

Also, earth's magnetic field suggests a 6,000-year time period since creation. Other planets do not demonstrate the same magnetic field that earth does. This field should have decayed over time and yet it still exists. This finding is consistent with a recent creation of this earth.[7]

The Young Universe/Young Earth view on the origin of life is consistent with that of the Old Universe/Hydroplate model, at least in most respects. In both models, God created all life on this planet, but did so just 6,000 years ago. In both, on Day 3 God created the plants; on Days 4 through 6—the birds, fish, land animals and, finally, humans were created. Neither model is dependent upon billions of years of suffering and death, or upon biochemical evolution and the endless speculations about how such complexity, function, and purpose could have arisen without intentional and conscious design.

Instead, God supernaturally created an endless variety of animals and plants, as well as Adam and Eve, with the genetic capacity for variation and adaptation to a changing environment, which—at least in the case of plants and animals—led to speciation over the following millennia. Advanced knowledge of genetics has strengthened this position, especially in the past few years. We now know that mitochondria in our bodies possesses its own DNA. This mitochondrial DNA is passed from mother to child—always on the X chromosome; that is, always from our mothers. Also, small genetic changes occur in our mitochondrial DNA in each generation. Thus, when we compare the various mitochondrial DNA sequences of the human race, and working with the known rates of variability, we can calculate how many generations back we have to go in order to reach our common ancestor. And, no, it wasn't 200,000-300,000 years, either. Instead, the journal Science (no friend of any creationist model), in 1997, gave us the answer—6,000 years to the common ancestor for all mitochondrial DNA.[8] Yes—six thousand years! More recent studies have been done with the Y chromosome for men, and similar findings have been seen for the male chromosome as well.[9]

The Young Universe/Young Earth and the Old Universe/Hydroplate views regarding the geologic column are the same. Both believe that the Bible provides an accurate, literal history of the worldwide flood depicted in Genesis; thus, the geologic column is a byproduct of this flood.

Geomorphology is interpreted similarly with these two literal creationist models. However, OUHP proposes additional elements that go beyond the scope of the traditional YUYE model as we shall demonstrate.

Old Universe/Hydroplate (OUHP)

OUHP theory assumes not only a Creator God, but One who is actively involved in His creation now—even if He first created the cosmos itself millions or even billions of years ago. He is a God who cares for His creation and who is conscious of the things that He has created. Those who adhere to this cosmogenic model believe that the universe itself had been created prior to the development of planet Earth and, therefore, has existed for billions of years prior to life on earth.

Belief that the universe itself has been around for billions of years also allows these Bible believers, but ancient cosmology theorists, flexibility when analyzing things like the "red shift." The red shift is the fact that light farthest from us on earth is found more on the red spectrum, or the longer wavelength spectrum, than is light closer to us. This phenomenon has been interpreted by secular scientists to suggest that the universe is expanding. The OUHP model agrees that an ancient Creator God stretched out the universe billions of years ago and continues to do so.

It goes on to suggest that God supernaturally established the earth long ago in the very distant past and there was no activity on this planet of water and rock—which was *tohu va vohu* ("without form and void")—until 6,000 years ago, when God began His life-on-earth construction project. This view, if correct, does allow for a dating of original granite rock on the planet being placed in the million to billions of years range, which we detect now, and which argues for the earth itself to, indeed, be "old." If there was radioactive decay of the original granite and other rocks during the long ages in the past when the *tohu va vohu* earth existed before the Genesis six-day creation started, then this may account for the long ages that radioactive dating gives us.

Young Earth Hydroplate theorists and other young earth creationists counter that there was no radioactivity prior to the flood and perhaps no radioactivity prior to the entrance of sin in the world. They suggest that radiation prior to the entrance of sin on this planet would not have been consistent with a God who created all things perfectly. Radiation prior to the entrance of sin would have been a source of birth defects and cancer in living beings, and God would not have allowed for this to occur.

Other scientific evidence in favor of a younger time frame for the *de novo* creation of earth includes helium diffusion rates in zircons and earth's magnetic fields. Helium diffuses rather rapidly from zircon crystals; thus the presence of helium in these crystals suggests their recent origin. The rate of decay of earth's magnetic field suggests a recent creation of the earth as well. Secular scientists postulate that the earth's magnetic fields undergo reversal every so often. Thus, when we see the magnetic field decaying, it is just moving toward the process of reversal. The mechanism for reversal is difficult to conceptualize.

The Old Earth Hydroplate model proponents would counter these arguments by pointing to the violence of the flood with the eruption of the fountains of the deep. According to this model, fusion reactions would have produced heavy elements deep under the crust of the earth. These heavy elements would have begun a radioactive decay process. Zircon crystals would have been produced at this time, accounting for the helium found in them. The earth's magnetic field would have been disturbed as well. This model has room in it for there to have been radioactivity decay of the original granite dating from perhaps billions of years ago, and it has room for those who suggest there was no radioactive decay prior to the entrance of sin, and thus that original granite did not decay prior to the flood and all radioactivity would then have developed as a consequence of the violence of the Hydroplate model of the flood.[10]

Critics of all the various young earth, global flood theorists point to a problem with heat production. At least three processes would have produced a massive amount of heat during the global flood. First, rapid movement of earth's granite plates during the global flood would have generated a massive amount of heat from the friction of these plates sliding—on water, per Hydroplate model, and on molten lava, per catastrophic plate tectonics. Second, young earth models suggest that a process of rapid radioactivity decay occurred during the flood, thus accounting for the appearance of "old" radioactive elements such as uranium/lead and potassium/argon ratios. This rapid radioactive decay would have produced a large amount of heat. Third, with the Hydroplate Theory, it is suggested that under the immense pressure of earth's plates, fusion reactions would have occurred in the subterranean water chambers increasing the potential energy and providing for the powerful eruptions of the fountains of the deep.

The Hydroplate model proposes that this third heat problem would have been addressed by a process called "cold packing" in which fusion reactions can occur without the release of massive amounts of heat. Work has been done in the Ukraine demonstrating this in their nuclear laboratories.[11] The other issues of heat generated as a result of friction (less with water than with magma), high water pressure, or rapid radioactive decay are dealt with by the projection of the massive amounts of water into the upper atmosphere where the heat could have been dissipated in this extremely large cold environment prior to the water returning to earth in the form of rain. Standard catastrophic plate tectonics does not provide a mechanism for this heat transfer as the water is not expelled to such great elevations and thus the heat is retained on the earth.

All this, though, is merely a precursor to what the real issues are. And they come down to the who, the what, and the why of life on earth.

Evidence abounds that God is the supernatural, intentional Creator of life. The complexity of life and its reproduction, adaptation abilities, and diversity scream of a Creator. Arguments including irreducible complexity, DNA replication and coding, protein synthesis, sexual reproduction, the lack of transitional life forms, the lack of pre-Cambrian evidence for life forms all speak as evidence of the Hand of God in creation.

OUHP and YUYE point to similar scientific information regarding the geologic column and geomorphology as well. We will next look at a few nuances of the Hydroplate model that further buttress its validity over that of catastrophic plate tectonics.

Unique to the Hydroplate Theory is the concept of liquefaction.[12] This is a process where water flows through layers of sediment in pulsations, and the layers are separated according to density, size, friction, and other physical qualities. This process may have caused organisms, all buried in a massive jumble, to separate more definitively according to these qualities, which explains why various animals that were similar to each other are found today at similar places in the geologic column. Massive amounts of mud and water in combination are necessary for liquefaction.

The massive mud flow from Mount St Helen's eruption demonstrates this process of rapid layering. Many mudflows occurred where the high-water content of the debris flows allowed for the dramatic sorting of soil layers as the mudflows came to a halt. As these fast-moving slurries lost energy, the sediment load was deposited in a sorted fashion, which is a classic characteristic of sedimentation in flowing water.

The massive size of the flood, and its association with enormous sedimentation, is necessary to account for the geologic column that we see today. Fossilization requires sudden burial in mud without exposure to oxygen, which promotes decay. The appearance in the fossil record of animals giving birth, eating other animals, and engaging in various other activities suggest that the animals were buried alive, unexpectedly, suddenly, massively. How else to explain fossils like one animal eating another? How could that possibly have been fossilized over long ages? It couldn't. It would be like leaving open the shutter of a camera that was supposed to take a picture instantly. The idea, then, that they were buried over millions of years—and with no erosion exposing the fossils and removing their remains—makes the slow accumulation of the geologic column unlikely. A massive cataclysmic flood is, by far, the most likely scientific event to account for the geologic column.

Mountains, for instance, are not all the same. Some are covered with marine animal fossils. Some mountains have steep slopes and sharp peaks and ridges; others show evidence of erosion. Most of those who advocate for an old earth postulate that the mountains were pushed up by the forces of plate tectonics over long, long ages. This motion is thought to be slow, inch by inch, year after year for millions of years.

Returning to our car crash analogy. If two cars were traveling at one another from a quarter mile away and accelerating to a high rate of speed, and hit head-on, a tremendous amount of folded metal and damage to the vehicle's shape will obviously follow. Now, for the kind of significant deformation to the land that we see with these mountain ranges, they must collide with a high momentum as well. That scenario best explains the contortion of the landscape. The force necessary to deform the mountains slowly over thousands to millions of years is beyond physical possibilities. In other words, it could not have happened. However, if the tectonic plates are moving rapidly, we would expect a tremendous uplift of the land, which is precisely what we see. To differentiate between plate tectonics and the Hydroplate Theory, we would then picture the land sliding on an extremely low viscosity layer of supercritical water, per the Hydroplate Theory, compared with a very high viscosity layer of lava associated with CPT. The evidence here, again, supports OUHP.

Additionally, the erosion patterns seen after the formation of mountains and valleys are telling. Erosion, which should have taken place on the surface of these mountains, also should have both taken away the fossils and shaved down their sharp mountain peaks and ridges as well. In short, water and wind should have—over the millions of years suggested by SS, TE, DA and AG—worn down these surfaces big-time. They didn't. There should not be the sharp peaks or mountain ridges that we see everywhere, and in abundance too, a fact that does not bode well for long and slow erosion. For example, formations like the edges of the Grand Canyon should be rounded, not sharply angled, that is, if millions of years have transpired since the Colorado river began to carve out the Canyon. All this means, then, that the evidence for a short time of erosion is strong.

Plateaus are difficult to explain as well, if one uses any long ages model. These are areas of the landscape that, according to their sedimentary layers, should be at a lower elevation than where they are found. However, some force had to work from underneath to push them up, like an elevator lifting them from over hundreds of square miles below. These areas are always adjacent to mountain ranges. The sedimentary layers are relatively undisturbed in their layering; however, they have been elevated, sometimes more than a thousand feet, from some massive force from beneath. If this force occurred slowly, over millions of years, one would expect weathering and channeling through the layers—but we do not see this. It appears that the plateau layers were elevated rapidly, and evenly. The Hydroplate Theory postulates[13] that these formations came from adjacent mountain ranges, which occurred from the compression event, weighing down on the underlying volcanic lava layers and producing a hydraulic lift elevation effect.

Again, the typography is best explained with the OUHP model.

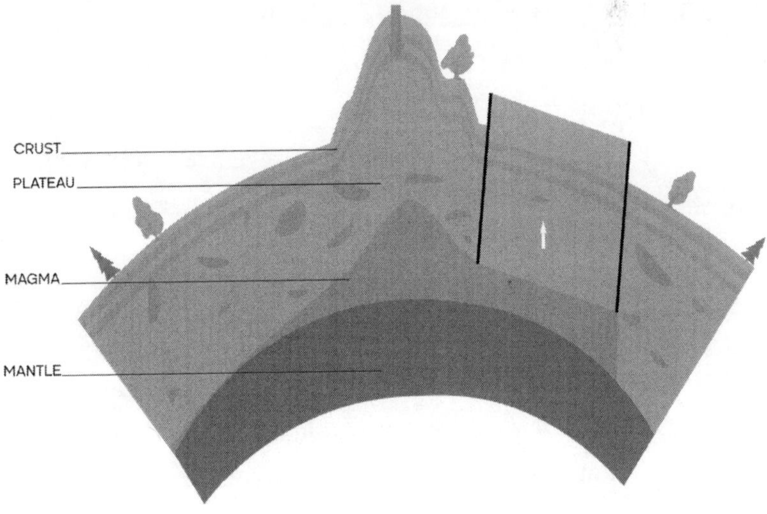

Limestone formation offers another interesting issue.[14] SS, TE, AG, and DA all look to ancient slow processes for formation of this rock. Limestone encases the majority of the carbon on the planet.

About 99.9% of the carbon is in limestone. Compare this with the extensive plant grown on the earth. All of the carbon in plant life and coal combined accounts for less than 0.1% of the carbon on planet earth. Thus, almost all the carbon is in the limestone. The theory goes that coral formed in shallow, warm seas and produced this carbonaceous deposit in the limestone. Because layers of limestone are, in places, tens of thousands of feet thick, this would have required a sea level variability of thousands of feet with an even temperature during the entire process over millions of years. This "just-so" scenario is extremely unlikely, to say the least.

A much more tenable explanation comes from the Hydroplate Theory. Carbonaceous sedimentation was produced as the subterraneous supercritical water interacted with the primordial rock. The carbonaceous sediment was expelled onto the surface of the earth, layering massively across continents and producing the extensive sedimentary limestone seen today. Old earth theories struggle to scientifically account for the limestone on the planet. Even the YUYE group do not have a clear explanation for where these thousands-of-feet-thick layers of limestone came from. Only the Hydroplate Theory offers a clear scientific explanation for the massive accumulation of carbon in the form of limestone on earth's surface.

Summary

Of the various theories of origins of our universe, of the earth, of life on it, and of its surface features, the scientific evidence favors an old universe, as well as an earth that was, likely, in existence as a granite rock covered with water prior to the biblical creation, perhaps here billions of years ago. We see evidence of a supernatural being—One we call "God"—who formed the earth, and then created life on it, a God who stayed involved in His creation from the beginning and afterwards.

There is strong evidence, too, that He created life on earth 6,000 years ago, and that—using the forces of gravity primarily as the mechanism—He had destroyed most life from the earth in a cataclysmic, worldwide flood known as Noah's Flood. Finally, the surface features of the earth, such as mountains, valleys, the water on earth's surface, the geologic column, and other geomorphic landscapes show evidence of this worldwide flood.

However, the Christian appeals to a Higher Authority than that of science: the Bible. Although the Bible requires, as does science, some interpretation, turning to Scripture gives us more authoritative evidence of the crucial truth about our origins, the Truth upon which all other truth rests. And to that source we turn next.

CHAPTER 5

Scripture

For precept must be upon precept, precept upon precept, line upon line, line upon line, Here a little, there a little. — ISAIAH 28:10

The Bible is a unique book, unlike anything else in the Western Canon, and its influence has been, and remains, immense. It makes claims about itself that, if false, are outrageously ludicrous; if correct, these claims present the truth of the universe. The Bible declares itself inspired by God, the Creator of the heavens and the earth. In many ways, it is self-authenticating, revealing things that, if openly and honestly investigated, prove its authenticity.

Many higher critical scholars—some claiming belief in God, others not (yet both groups often working from modern, naturalistic presuppositions)—seek to denude the Bible of its divine character, attributing much of what it says to mere human ideas about God. All the Bible writers were, they assert, heavily influenced by culture. The higher critics also claim that many crucial biblical accounts, such as the Genesis creation, are allegory or legend. They capitulate to the secular critics and acknowledge the impossibility of miracles, or prophecy, or the supernatural. But bringing such naturalistic presuppositions to the Bible makes about as much sense as using a French dictionary to interpret a Chinese text.

Rejecting this approach, and assuming the Bible is indeed the Word of God, and that it should be taken as literally as possible, we will look at our six categories of origin theories, and seek to determine how well, or poorly, they harmonize with the Bible.

Secular Science

The secular scientist basically looks at the Bible as a piece of literature written by Jews (mostly) between 2,000 and 4,000 years ago, all in an attempt to validate their religion. In this view, religion in general is deemed merely a mechanism of regressive human thought, a primitive attempt to explain and, to a certain degree, cope with their surroundings. One of the most revolutionary differences, however, between the Bible and other ancient religious texts is that instead of ascribing deity to natural phenomena like the sun, the moon, and the stars (planets)—the Bible writers saw them as the inanimate creations of an omnipotent God. Nevertheless, for these critics, the biblical creation and flood narratives were just other ancient legends similar, for instance, to the *Enuma Elish,* the main creation account of the ancient Babylonians. However much some of these secular scientists may respect the Bible (or may not) as literature, none take the first 12 chapters of Genesis seriously, or at least as any accurate account of origins, or of evil.

In other words, the Bible has very little, if any, impact on secular science's theories of origin, except perhaps to be used as a foil to be spoken against, and to be avoided at all costs. If the Bible suggests one thing, most secular scientists will bend over backwards to avoid likeness to it. By default, then—and based exclusively on a priori philosophical assumptions, and certainly not on the amazing beauty and complexity of life—this theory denies the possibility of any Intelligent Designer in the origin of the universe or of life.

Theistic Evolution

Theistic evolution opens the door to Genesis 1:1—"In the beginning God created the heavens and the earth." Adherents usually view this verse as literal. They are open to a God providing the initial supernatural intervention that established all matter. For without something outside of nature, even greater and transcendent to it, how could "nature" itself (a category of reality) have arisen in the first place? This conundrum is something that atheists have a hard time logically addressing.

However, while accepting the Big Bang as their model for the initial creation event, most theistic evolutionists read Scripture in the fullest sense as do liberal historical-critical theologians. All Scripture after Genesis 1:1 is an allegory—at least the part discussing how life arose on the earth. They, in fact, tend to be very skeptical of any miracles in Scripture, not only the miracles involved in the creation, either. These "miracles" include the parting of the Red Sea, resurrections of the dead (including Christ's), or prophecy having any predictive power, and so on. Thus, to spend time comparing the various theistic evolutionary theories floating out there to the Bible is pointless. Many theistic evolutionists don't expect the Bible to support their theory; those who do tend to twist and distort the texts in order to try and fit them with evolution that it's not worth the effort to refute them.

Active Gap

The Active Gap group reads Genesis 1:1 as a summary of the billions of years needed (they believe) for the development of the universe in general and of the earth in particular. Along with this development was the creation of animals, plants—and apes, which, over long ages, evolved into human-like animals, the precursors to *Homo Sapiens.* There is, though, no scriptural support for this view; in fact, everything in the early chapters of Genesis militates against this idea.

Why, then, do they argue for this? It is because they are swept up in the basic secular view of reality, with its dominant understanding of the geologic column, which teaches longs ages for the development of life on earth. For them, to deny science is to deny logic and reason. Thus, they fall back on a concocted story of the planet and its history, all in order to stay aligned with the secular view of cosmology and geology. They then turn to Genesis 1 and 2, and argue that God breathed into an evolved early human His Spirit; and, thus, He created the first Adam. Up to this point, we see no correlation with any honest reading of Scripture and this view.

Nor do Active Gap theorists accept a global flood either. The geologic column is attributed to eons of time prior to the second creation of Adam; and, thus there is no need for a global flood. Rather, this group views the flood in Genesis 8 as a local event that impacted the early descendants of Adam profoundly enough for them to develop a legend around that local flood event. Real history begins, for them, perhaps at the tower of Babel. However, their allegorical approach to the first 10 chapters of Genesis causes this group to also view many other portions of Scripture with the same historical critical approach. This approach can only weaken, at least in the minds of those who accept it, such biblical truths as predictive prophecy and the key events regarding the ministry of Jesus on earth.

Day-Age

The Day-Age view is the first to really claim a commitment to believing in Scripture as it is written. But does it accurately interpret what Scripture teaches?

Genesis 1:1 reads: "In the beginning God created the heavens and the earth." This is understood as a title statement of what is going to follow, which is pretty much how the Young Universe/Young Earth creationists view it as well; however, they differ widely in their interpretations from this point on.

Genesis 1:3, 5 reads: "Then God said, 'Let there be light'; and there was light.…So the evening and the morning were the first day." Here the Day-Agers interpret the word "day" as an "age," perhaps a billion years long or even longer. There is no specific explanation of what was happening during this "day" of creation. It may represent the formation of energy, or the Big Bang, or the evolution of the elements. Who knows? The important point for this discussion is that this "day" is not a 24-hour period. It is the first epoch of creation, when God used His creative power to establish electromagnetic energy in the universe, however long that supposedly took.

These long "days" of creation in the first two chapters of the Bible are also the time periods during which, they propose, the geologic column formed. On each successive "day" (really billions of years), more and more accumulation of sediment developed, particularly during the "days" of fish and birds (Day 5) and the land animals (Day 6). Animals, evolving over these eons, died, their fossils in the layers of earth being interpreted as the evidence for evolution. By the time of Adam and Eve, the entire geologic column would have existed.

One rather glaring problem with this view (if one takes the Scriptures seriously, as the DA folks claim that they do), is that the Scriptures clearly read: "and the evening and the morning were the first day." The "evening and the morning," just like a 24-hour day now, was composed of, well, yes, an evening and a morning. Why would the inspired record so succinctly depict it that way if it really didn't mean an evening and a morning but, instead, millions or billions of years? The DA group, however, holds that "science" insists on this vast amount of time, and, because they give science the ultimate say, then it must mean an epoch of time, and not just 24 hours.

This theory, with each successive day representing millions or billions of years, leads to another problem—especially during the "days" in which life appears—because each day is described in Genesis as being "good." But how could each day be "good" if each "day" were filled with the bitter, brutal, and violent massacre known as the "survival of the fittest," which a loving God used to create life as we now know it? (And the problem of the lack of the sun for light has previously also been discussed).

And what about God resting on the seventh day? No problem. They argue that we are still living in the seventh day of creation, in which God is resting from the work of His creation. But Genesis 2:3 says that God "blessed the seventh day and sanctified it." Is it possible that God blessed and sanctified the seventh day, yet this "day," in which we live now, is filled with sin, suffering, death, and destruction?

Worse, if the Day-Age Theory were correct, the fall in chapter 3 is almost pointless. The earth has already experienced death and destruction, billions of years of death and destruction. But death is the result of sin. Romans 5 makes that perfectly clear. So, for instance, how is Cain murdering Abel any different from the alleged early hominids killing one another as the process of survival of the fittest ape-man developed over eons of time?

Also, the Day-Age theorist is forced to take a historical critical view of the next nine chapters of Genesis, since they argue that these long epochs of time were the mechanism for the Grand Canyon, dinosaur and other fossils, as well as the various landforms and ice ages of our planet. It is only after Chapter 11 of Genesis that they can revert back to a literal reading of Scripture and change to a more conservative view of the Bible. This rather arbitrary back and forth between literal and symbolic, between higher critical and literal, undermines the entire validity of Scripture.

Young Universe/Young Earth

The final two views, Young Universe/Young Earth (YUYE) and Old Universe/Hydroplate (OUHP), both claim adherence to a literal reading of the Bible; this, in contrast to the views already discussed above, which either discard the Scriptures entirely or that view the early creation and flood accounts as symbolic, allegory, legend even. Thus, for the conservative Bible student, the argument really comes down to how these last two views hold up against Scripture.

Genesis 1:1 states, "In the beginning God created the heavens and the earth." Now, at this point in the narrative, nothing has been stated about the creation of water, earth, light, etc.—all that was yet to come in the six-day creation narrative. Therefore, the YUYE creationist views verse 1as a title statement. To paraphrase the first verse, a young earth creationist might say something like: "In the beginning of all things, God created everything in the universe including the earth and its solar system, and what follows are the details of how God created those things."

Genesis 1:2 reads: "The earth was without form, and void; and darkness was on the face of the deep. And the Spirit of God was hovering over the face of the waters." Again, up to this point, nothing has been said about the creation of any granite mass, rock, or form of the material that was to become the earth. The earth is depicted as already being there. The same with the water: it's just there. Thus, the YUYE creationist also views verse 2 as a title verse. They might paraphrase the verse in this manner: "The earth was nonexistent, and all was dark in the universe. God the Father, Son, and Holy Spirit came to the location where the earth and its components were going to be created."

However, this view brings a big problem to the YUYE theory: the water and earth are there already. For something to be "without form and void" means that, whatever it is, yes, it is empty—but it exists! The statement that "the bucket is empty" implies a bucket. Where, then, did this mass and water come from because it had, obviously, been created before Day 1?

Genesis 1:3-5, "Then God said, 'Let there be light'; and there was light. And God saw the light, that it was good; and God divided the light from the darkness. God called the light Day, and the darkness He called Night. So, the evening and the morning were the first day." The YUYE view sees this verse as the initiation of all created matter in the universe. (After all, Einstein showed, with e=mc2, that matter and energy were different forms of the same thing.) Electromagnetic radiation is now established in the universe.

Now it is important to note that the stars were not created on this day, nor the sun, but only light and energy. The day/night cycle is established without the sun in the picture. Night is simply the absence of light, and day is simply the presence of light. The sun does not have to be the only source of light, and so it is not necessary for a day/night cycle. The evening and the morning were the first day and, thus, one rotation of the planet, a 24-hour period. With the YUYE view, there is no consensus on whether the earth, its mass and elements, were created on this day. With this view, it seems most apparent that the earth was created on the first day when light was created because, for an evening and morning to transpire, there must have been a rotating planet.

Genesis 1:6-8, "Then God said, 'Let there be a firmament in the midst of the waters, and let it divide the waters from the waters.' Thus God made the firmament, and divided the waters which were under the firmament from the waters which were above the firmament; and it was so. And God called the firmament Heaven. So the evening and the morning were the second day." For the YUYE theorist, these verses describe the creation of the atmosphere. Likely, the establishment of oxygen and air pressure. This day may also describe the creation of Heaven, where God lives, at least according to this view. As Paul calls it, the "third Heaven." Up to this point, there is no sign of the sun, moon, or stars. There is no evidence of any other beings such as angels – only light, earth, water, and atmosphere.

Genesis 1:9-13,

> "Then God said, 'Let the waters under the heavens be gathered together into one place, and let the dry land appear'; and it was so. And God called the dry land Earth, and the gathering together of the waters He called the Seas. And God saw that it was good. Then God said, 'Let the earth bring forth grass, the herb that yields seed, and the fruit tree that yields fruit, according to its kind, whose seed is in itself, on the earth'; and it was so. And the earth brought forth grass, the herb that yields seed according to its kind, and the tree that yields fruit, whose seed is in itself according to its kind. And

God saw that it was good. So the evening and the morning were the third day."

The YUYE view read this as follows: "God raised the land mass above the water in some places and sunk the waters deeper in other places, all in order to establish land and sea. He then created plants and biology, with the ability to reproduce and to provide food for His coming created humans and animals."

Genesis 1:14-19, "Then God said, 'Let there be lights in the firmament of the heavens to divide the day from the night; and let them be for signs and seasons, and for days and years; and let them be for lights in the firmament of the heaves to give light on the earth'; and it was so. Then God made two great lights: the greater light to rule the day, and the lesser light to rule the night. He made the stars also. God set them in the firmament of the heavens to give light on the earth, and to rule over the day and over the night, and to divide the light from the darkness, and God saw that it was good. So the evening and the morning were the fourth day." Here, the YUYE creationist views this as the establishment of the entire universe. The sun, moon, planets, our galaxy, and the two trillion galaxies in the universe were all created during this 24-hour period. Some believe heaven, God's place of dwelling, was also created on this day, along with the angels (some place angels on Day 6, with the creation of man).

Days 5 and 6 see the creation of water, air, the animals, as well as Adam and Eve. He created them male and female with the ability to reproduce, to diversify, and to adapt.

Genesis 2:1 states, "Thus the heavens and the earth, and all the host of them, were finished." This is an important verse for the YUYE view as it closes the work of creation. What was started in Genesis 1:1, the creation of all things, which means "the heavens (universe) and the earth," was finished in verse 2:1. This gives credibility, in their minds at least, to the idea that Genesis 1 is talking about the entire universe and not just this planet.

The remainder of Genesis 2 is a recapitulation of the creation story of chapter 1, but with more details added. It is in chapter 3 with the fall that more divergence of views exists.

Genesis 3:1, "Now the serpent was more cunning than any beast of the field." We know from Revelation 12:9 that this serpent was either the devil himself or had simply taken over the serpent. The serpent by itself could not have been wicked because God made all of His creation perfect. If the devil had infiltrated this planet, even Eden itself, and assumed the form of one of the animals (or possessed it)—it begs the question, *Where did the devil come from? How long has he been in existence? Was he part of God's perfect creation mentioned in chapter 1 of Genesis?* In Revelation 12, there is mention that the devil deceived a third of the angels in heaven and they were cast out of heaven. *When did this occur? How long did it take for the devil to deceive the angels in heaven?* Revelation 12 discusses a "war" in heaven. How long was this war? Who was involved in it?

The YUYE model goes into crisis mode over Revelation 12 and Genesis 3. How could they not? Surely, only a short time had taken place between the creation story and the fall in

Genesis 3. There is no mention of any children being born prior to Genesis 3. Adam and Eve, being made perfect, were most likely very fertile. Having been told by God to be "fruitful and multiply" (Genesis 1:28), Adam and Eve, created as adults in full maturity, would have had children a short time after their creation. This provides a very short time period from the creation of everything in the universe to the fall of Adam, during which time Satan fell and created a war in heaven. How could Satan have had enough time to become corrupt, to sin, to deceive a third of the angels, and then be cast out of heaven to the earth? Again, although we don't know how long after creation the fall came, it does not seem to have been long enough for all that happened with Satan in heaven to have occurred.

The antediluvian history of the earth continues, with the two views sharing, by and large, a common understanding of the earth and its inhabitants; that is, until the flood.

Genesis 7:11, "In the six hundredth year of Noah's life, in the second month, the seventeenth day of the month, on that day all the fountains of the great deep were broken up and the windows of heaven were opened." The YUYE model proposes a mechanism for the flood based on the immediate and rapid movement of the continental plates, called Catastrophic Plate Tectonics (CPT). We have discussed this in prior chapters. The fountains of the deep here in verse 11 are the lava flows that would have resulted from the rapid movement of the continental plates. This lava flow would have initiated movement of the plates, with a resulting tsunami effect on earth's seas. These actions, combined with the heating of the seas, led to the evaporation of water, resulting in tremendous rain. This is the proposed mechanism for the entire earth to be flooded, even above the highest mountains. If the same volume of water was on the surface of the earth at the beginning of the flood as was on it at the end, the sea floors would have had to rise significantly, in order to force the flooding of the landmass—particularly, as Genesis 7:19 states, above "the high hills." Although CPT theory involves plate movement, the net elevation changes required for flooding of the entire continental landmass is not adequately accounted for, at least without the addition of water from outside of the surface of the earth, which makes the whole theory, basically, untenable.

Old Universe/ Hydroplate Theory

Genesis 1:1—"In the beginning God created the heavens and the earth." Those who believe in the Bible and an old universe can look at this verse in two ways. One is to consider that the heavens and the earth that are mentioned in verse 1 are the sky, the local space in our solar system, and the atmosphere—and not the distant heavens of the entire universe, or the heavens where God dwells. This line of thinking suggests that Genesis is talking just about the creation of our planet, because, after all, this is the topic at hand. To paraphrase, "At the beginning of this world, God created the sky and the earth."

The other way of looking at this verse is to consider that the word "beginning" may be referencing not just the beginning of this world, but the beginning of all things—much the same way YUYE creationists view this verse. However, at the end of this verse, many old universe literal creationists would place a long gap of time. This view is called the "passive gap" theory. A long gap of nothing happening on the earth for, perhaps, billions of years until the events of verse 2 arrive. This view would be paraphrased as, "In the beginning of all things in the universe, God created the entire universe, all the stars and galaxies seen and even placed the earth out there in space—a rock of granite covered in water. After, perhaps, billions of years, God decided to complete the creation of the earth, with all sorts of life forms, including, even culminating with, human life." Verse 2 is where the beginning of this process is first depicted in Scripture. Both of these approaches work for the old universe theorist.

Now, believers know that God is everlasting and eternal: "Have you not known? Have you not heard? The everlasting God, the Lord, the Creator of the ends of the earth..." (Isaiah 40:28). "Now to the King eternal, immortal, invisible, to God who alone is wise ..." (1 Timothy 1:17). "Your throne is established from of old; You are from everlasting..." (Psalms 93:2) Thus, a connection is made between God being eternal and everlasting with the Bible phrase "of old." When in Scripture the term "of old" is used in reference to God's throne and to His Creatorship, we must consider that this phrase may refer to "eternal" or "everlasting" time as well—and not just back to the beginning of our world. Surely God is much older than just 6,000 years, as is His throne. The following verses all make reference to God as being the Creator "of old" (Psalm 68:33; Psalm 93:2; Psalm 102:25; 2 Peter 3:5).

The next line of evidence for the old universe is that Scripture describes God as "stretching out" the heavens. This stretching out of the universe appears to be consistent with the observed "red shift" that astronomers see in distant galaxies—a continuous expansion of the cosmos. The following verses make reference to God stretching out the heavens: Psalm 104:2; Isaiah 40:22; 42:5; 44:24; 45:12; 51:13; Jeremiah 10:12—all powerful evidence for an old universe.

Additionally, the Bible makes reference to "laying the foundations of the earth." This would be an event that occurred prior to God creating light (Genesis 1:3). The phrase, "laying the foundations of the earth," appears to reference God creating the solid granite planet and the water on it, over which He hovered (Genesis 1:2) at some distant time in the past prior to the creation story as depicted in Genesis 1:3 onward. The verses that make reference to the laying of the foundations of earth are Job 38:4; Psalm 104:5; Isaiah 51:13.

Old universe literal creationists also point to things mentioned in the Bible that are not included in the six days of creation in Genesis 1 and 2. These include:

- Details of the stars: Job 9:9; 38:32; Psalm 147:4; Amos 5:8
- The angels and Lucifer: Isaiah 14:12; Revelation 12:7
- Invisible things: Colossians 1:16 says God created "invisible" things. Again, there is no mention of these in Genesis.
- The host of heaven: Jeremiah 33:22 says the host of heaven cannot be numbered. Who are they? They are not mentioned as being created in Genesis 1 and 2. Perhaps, angels?

- The sons of God in Job: Job 1:6; 2:1; 38:7. Who were these sons of God? Were they angels? Were they other created beings? Nothing is mentioned of them in Genesis 1 and 2.
- The morning stars mentioned in Job 38:7 "sang together" when God created the world. Were these angels? They must have been created prior to this earth, then, whoever they were?
- Principalities and powers in heavenly places: The Bible is filled with language about other life in the cosmos. Ephesians 3:10; 6:12; Colossians 1:16; 2:15; Titus 3:1; 1 Peter 3:22. Were all these created only within the past 6,000 years, and only after the creation of earth?
- Worlds: Hebrews 1:1-2 says, "God, who at various times and in various ways spoke in time past to the fathers by the prophets, has in these last days spoken to us by His Son, whom He has appointed heir of all things, through whom also He made the worlds." This same Greek word, aion, translated "worlds," is in its plural form. The same word is used in Hebrews 11:3—worlds. If the Godhead made "worlds," what other worlds are being referred to here besides the earth? The writer of Hebrews appears to be implying that God made other sentient beings in other "worlds."

Genesis 1:2 reads: "The earth was without form, and void; and darkness was on the face of the deep. And the Spirit of God was hovering over the face of the waters." This verse is strikingly similar to Jeremiah 4:23 where the prophet states, "I beheld the earth, and indeed it was without form, and void;

And the heavens, they had no light." This passage in Jeremiah appears to be making reference to the Day of Judgment, when God will redeem the righteous and punish the wicked. Is there a connection between the form of the earth prior to creation and the state of the earth after God comes to judge the earth at the second coming? Many believe that is what these verses are saying, and with good reasons too. Revelation 20:1-3 describes this time period when the devil, Satan, will be bound to this earth in the "bottomless pit." Comparing this passage with Jeremiah 4, we can picture this earth returning to its "without form and void" state during the thousand years when the devil is bound to this lifeless, lightless planet. That is, the state of this earth prior to Genesis 1:3 when God arrived here.

Genesis 1:3, "And God said, 'Let there be light'; and there was light." God's presence on this planet established light. The first day occurred as the planet rotated one complete motion, with God remaining over its surface as its source of light—creating the "day," and His absence on the other side of the earth creating the "night."

The day in question, the fourth day, goes like this:

Genesis 1:14-18 reads, "Then God said, 'Let there be lights in the firmament of the heavens to divide the day from the night; and let them be for signs, and for seasons, and for days, and years; and let them be for lights in the firmament of the heavens to give light upon the earth'; and it was so. Then God made two great lights: the greater light to rule the day, and the lesser light to rule the night. He made the stars also. God set them in the firmament of the heavens to give light on the earth,

> and to rule over the day and over the night, and to divide the light from the darkness. And God saw that it was good."

The beginning of the passage described the sun and moon being created and put in place. Toward the end, a separate sentence states, "He made the stars also." Old Universe adherents believe this is stating that He made the stars at a separate time and place. After all, this is the same God who made the stars and all of the other things not mentioned in the Genesis creation account, including the angels, other worlds, heaven, and everything else in the remainder of the cosmos. That is to say, this is the God who also made the stars and everything else besides the sun, moon, the solar system, and this earth.

How does one determine how Moses intended for this verse to be understood? "He made the stars also." Does this mean that He made them immediately after the sun and moon? Or does it mean that He made the stars at another time, at a different creation event? One must look at other Scripture to determine the answer.

Although the Bible may not offer certainty in terms of clarifying the phrase, "he made the stars also," other Scripture may shed light here. Psalm 147:4 reads: "He counts the number of the stars; He calls them all by name." If God created all the stars on Day 4, how could He call them all by name? This verse seems to imply that He created each star with special care and attention. Perhaps He named all of the 2 trillion stars in the universe in less than 24 hours. Certainly, God could do this, but it seems implausible. Job 38:7—"When the morning stars sang together, and all the sons of God shouted for joy."

This verse refers to the angels and other created beings shouting for joy at God's creation of the earth. These beings and the places they inhabited must have been created prior to the creation of the earth.

Isaiah 14:13, "For you have said in your heart; 'I will ascend into heaven, I will exalt my throne above the stars of God; I will also sit on the mount of the congregation on the farthest sides of the north.'" This verse is speaking of Lucifer rebelling against God. When was Lucifer created? It must have been prior to the creation of the earth. He had to have had time to be created, to be exalted (in his own mind), to sin, to rebel, and to deceive the other angels. This verse suggests that Lucifer wanted to exalt his throne above the stars of God. The stars must have been created prior to Lucifer rebelling. If the stars were created on Day 4, when did Lucifer rebel? It would have had to be later than Day 4. If Lucifer was created on the same day that man was created, why doesn't the Bible mention this fact? Instead, these verses about the fall of Lucifer show that the stars were already there before the six-day creation.

Genesis 2:4, "This is the history of the heavens and the earth when they were created, in the day that the Lord God made the earth and the heavens." Here, as in other verses in the first two chapters of Genesis, "heavens" refers to the sky and the local heavens of the solar system. Deuteronomy 1:28 says that "the cities are great and fortified up to heaven." This was the report of the spies on returning from Canaan. Certainly, the words mean that the city walls reached up into the sky; that they were high up. Second Samuel 18:9 reads: "So he was left hanging between heaven and earth"; this is in reference to

Absalom being caught by his hair in a tree. "Heavens" again refers to just the sky. Acts 2:19 says: "I will show wonders in heaven above and signs in the earth beneath: blood and fire and vapor of smoke." This is a reference to signs being shown in the visible sky above the earth. Depending upon the context, "heavens" can refer to the sky, or any space above earth's surface, or space extending from our atmosphere to our solar system and all the way out to the farthest galaxy. Finally, the third heaven as Paul refers to in 2 Corinthians 12:2 is the location where God's throne resides. Context is everything.

Genesis 2:5-6, "For the Lord God had not caused it to rain on the earth, and there was no man to till the ground; but a mist went up from the earth and watered the whole face of the ground." If there was no rain on the earth, there must not have been any large oceans. (This is what produces our weather patterns now.) With no rain, the "mist" or dew must have been heavy enough to water plants adequately. If the earth were warm, and the air cooled at night, the warm earth may have heated the rivers and seas, causing the water to evaporate and condense at night on the earth's surface, producing the morning dew. Hydroplate Theory provides a mechanism for such a heating of earth's crust, which could create the dew as previously discussed.

Genesis 3 refers to the fall of mankind. If Satan had existed in ages past in heaven, as Isaiah 14 and Ezekiel 28 suggest, he would have had time for maturing into the covering cherub that guarded the ark of God in heaven. He would have had time to have fallen from his lofty position, as pride crept into his character. He would have had time to deceive and entice a third of the angels in heaven, which could be in the millions. Time would have been allotted for some type of war in heaven between God and Satan, resulting in the devil with his angels being cast out of heaven. Despite being cast out of heaven, there appears to have been some sort of agreement between God and Satan to allow the devil to have access to the earth, but limited to a confined space on a certain tree in the Garden of Eden. This led to the young Eve—probably having experienced only weeks of life—being deceived and then, along with Adam, expelled from the Garden of Eden.

At the same time, the Bible supplies ample support for the Hydroplate Theory. To begin, water must have collected somewhere under earth's crust prior to and subsequent to the flood, as evidenced by the following verses: Psalm 24:1-2 says, "The earth is the Lord's and all its fullness, the world and those who dwell therein. For He has founded it upon the seas, and established it upon the waters." Psalm 33:7 reads: "He gathers the waters of the sea together as a heap; He lays up the deep in storehouses." Psalm 136:6 reads: "To Him who laid out the earth above the waters…." Second Peter 3:5 states: "For this they willfully forget: that by the word of God the heavens were of old, and the earth standing out of water and in the water…" (emphasis mine).

That water, which was under earth's crust, was expelled upward at the flood: Genesis 7:11 says that "…on that day all the fountains of the great deep were broken up, and the windows of heaven were opened." Job 38:8 says: "Or who shut in the sea with doors, when it burst forth and issued from the womb."

Proverbs 3:20, "By His knowledge the depths were broken up, and clouds drop down the dew."

Geysers and deep ocean water jets called "black smokers" are evidence for deep crustal superheated water. Job 38:16, "Have you entered the springs of the sea? Or have you walked

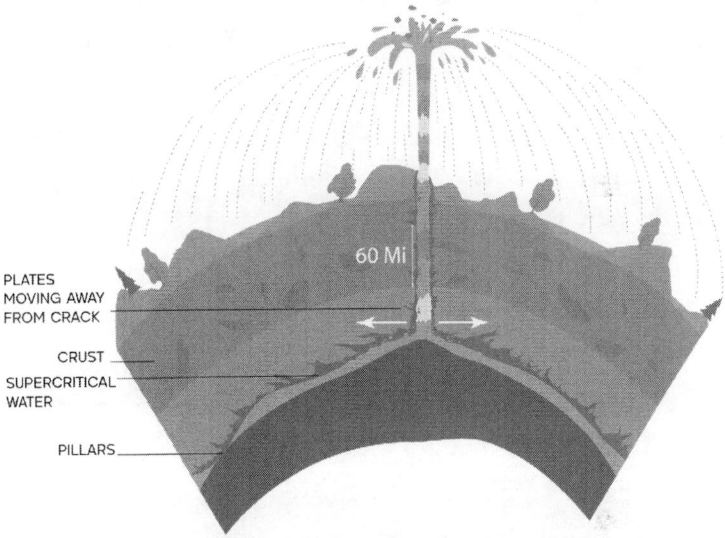

Initial 40 days of the flood, when rain was occurring.

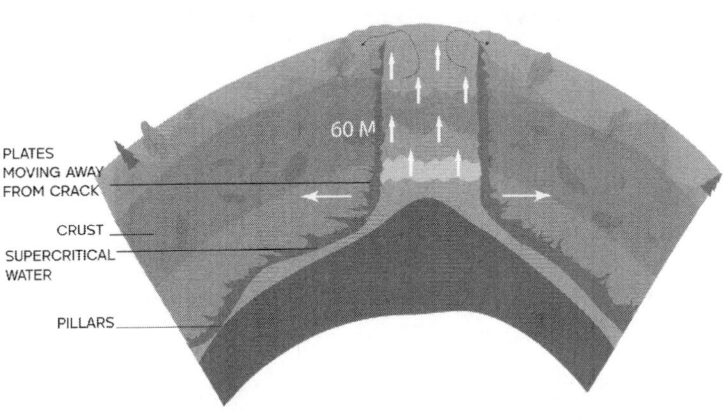

The water continued to increase for 110 more days after the rain ceased.

in search of the depths?" Could these "black smokers" be these "springs of the sea"?

The biblical account of the flood records the rain duration being for forty days, as revealed in Genesis 7:12. The water, however, continued to increase in depth for 150 days (Genesis 7:24; 8:3). This is consistent with the water being expelled from the earth, all the while rain was falling for 40 days; however, as the water depth increased on earth's surface, the expelled water was no longer launched into the atmosphere but, rather, continued to raise the water level until day 150—five months after the rain began!

The flood, therefore, was so much more than just heavy rain. It involved crustal plate movements, water shooting up from below the surface, volcanic activity, tsunamis, hail—and massive erosion and mountain formation *in a matter of hours.* Psalm 18:7-15 describes this: "The earth shook and trembled; the foundations of the hills also quaked and were shaken, because He was angry. Smoke went up from His nostrils, and devouring fire from His mouth; coals were kindled by it. He bowed the heavens also, and came down with darkness under His feet. And He rode upon a cherub, and flew; He flew upon the wings of the wind. He made darkness His secret place; His canopy around Him was dark waters and thick clouds of the skies. From the brightness before Him, His thick clouds passed with hailstones and coals of fire. The Lord thundered from heaven, and the Most High uttered His voice, hailstones and coals of fire. He sent out His arrows and scattered the foe, lightnings in abundance, and He vanquished them. Then the channels of the sea were seen, the foundations of the world were uncovered at Your rebuke, O Lord, at the blast of the breath of Your nostrils.'"

Finally, as we consider the evidence for the Hydroplate model as it relates to Scripture, let us turn to the second day of creation. Genesis 1:6-8 states, "Then God said, 'Let there be a firmament in the midst of the waters, and let it divide the waters from the waters.' Thus, God made the firmament, and divided the waters which were under the firmament from the waters which were above the firmament; and it was so. And God called the firmament Heaven. So the evening and the morning were the second day." The traditional way this is interpreted is that the firmament in this passage is, as stated, "heaven," and heaven is the atmosphere or the sky. Thus, the water vapor in the atmosphere, such as the clouds, is the water above the firmament, and the water which is in the sea is the water below the firmament.

The potential problem with this reading is that there is proportionately much less water vapor in the sky than on the earth. There was no rain prior to the flood; thus, the amount of water vapor in the sky may not have been that great. Furthermore, one might question why God would have dedicated a whole day of creation to make the clouds. Couldn't clouds have been made simply from the sun shining down on the earth, creating evaporation of the water and allowing it to rise up into the sky as clouds? Or, perhaps, the emphasis of this day was on creating oxygen? Yet, at first glance, it seems a small accomplishment for God, especially when comparing Day 2 with the other days of creation and what was achieved on each one.

The alternative view, which is consistent with Hydroplate Theory, but not imperative for the validity of the theory, is as follows: The firmament is stated in verse 8 to be heaven. Perhaps the earth, the Garden of Eden, was the place where God dwelt with mankind—and this was considered "heaven"? Thus, the firmament divided the surface water from the deep subterranean water, which was placed there for its tidal pumping effect to warm the surface of the earth (as stated in previous chapters). The word for "firmament" is raqia, which is translated as "solid supporting waters above." This view would seem consistent with the passage: The separation of the water above from below would involve about a 50/50 split of the water and would require significant rearrangement of the surface of the earth—consistent with the work of a Day of creation. The Hydroplate Theory is not dependent on this interpretation, admittedly speculative; however, if correct, this interpretation provides further support to the theory and mechanism by which God eventually brought the flood upon the earth.

Conclusion

For the Christian, the Bible should be the ultimate authority. Unfortunately, many Christians have relegated Scripture to the annals of mythology, tradition, and folklore. This view makes them pretty closed-minded to any kind of literal reading of the Genesis creation, which takes the biblical account literally.

Once a literal reading of the Bible is entertained, however, it is a matter of interpretation. We have presented evidence for and against a Young Universe/Young Earth creationist view and an Old Universe/Hydroplate Theory view. Both claim to be consistent with Scripture.

The YUYE view emphasizes that God began creating all things just 6,000 years ago and that the global flood was secondary to rapid movement of tectonic plates over the underlying mantle, with resulting tsunamis that flooded the world. The OUHP view postulates an eternal God who has been in the business of creation for billions of years, but began creating the earth and the surrounding solar system just 6,000 years ago. The flood occurred when the water, which God placed under earth's crust at creation, burst forth from its subterranean chambers onto the surface of the earth, flooding the land and reshaping earth's surface. There is biblical support for both; the reader has to decide, however, which one is the most consistent with Scripture.

In the next chapter we will look at the implications of the OUHP view. We believe that this view completes the puzzle and provides a coherent picture of the God of the universe.

CHAPTER 6

Implications

God, who at various times and in various ways spoke in time past to the fathers by the prophets, has in these last days spoken to us by His Son, whom He has appointed heir of all things, through whom also He made the worlds.

— HEBREWS 1:1-2

Like a bad argument between spouses, does it really matter who is right about origins? Or about how old the universe really is? Or about how the earth was formed? Or if the flood was local or global? Or how God brought the flood about in the first place?

How much do these questions really matter to how we live our lives, now, today?

They do matter, because these questions can impact how one relates to God, and how we understand the way that God relates to us. And, too, they impact the biggest questions of all:

Who are we, and what may we hope for in a world that, in and of itself, offers so little hope?

In this chapter, we will seek to demonstrate critical, universal issues as they relate to the character of God, to the depth of His sacrifice for mankind, and to His love for His creation, all of which are tied to the question of how He created us and the world to begin with.

The Love Thing

God is love. It sounds trite but it is a foundational truth of the Christian faith. But what does this mean for His universe? For instance, does God express His love in a vacuum? To love means to have an object to love, does it not? So, does He not need other beings to express that love to? This is partially answered with the concept of the Trinity—the doctrine that God is three persons in One: God the Father, God the Son, and God the Holy Spirit. Within an eternally existing Godhead, the three unified natures of God can express the quality of love between them.

How exactly love is expressed among the Deity is not easy for us, as mere mortals, to understand. However, as beings made "in His image," and with some capacity in ourselves to love, we can, in human terms, understand love as expressed in different ways, at least between humans. These ways include selflessness, unconditional sacrifice for others, empathy, patience, kindness, humility, encouraging others, bearing other's burdens, and so forth. None of these expressions of love make sense apart from there being other people to interact with and other people to love.

Another quality might include the desire to propagate love. This would involve God creating more beings to whom He could express His love. God is the Creator, the One who made "the heavens and the earth." Whatever once didn't exist but then came into existence, did so only through God, who is also Jesus (See John 1:1-3). An eternal Creator God who existed from eternity past, and who also is love—He surely had not waited until only six thousand years ago before creating a planet and beings on it whom He could love, especially in a creation billions of light-years in size, would He? So, very likely He had been creating other beings to love—and had been doing so for longer than our very temporal minds can fathom.

If this is the case, is it any wonder that, as we look out into the cosmos, we see evidence of vast time frames with aspects of reality difficult for us to explain with our limited concepts of physics and chemistry? Of course not. In fact, what else would we expect to see? Galaxy after galaxy, in numbers that stagger our mind, exist—and each comes with as much variability as we see in every different snowflake. That's what's out there in the cosmos that we—on this tiny planet hurtling through space at thousands of miles an hour and in all directions too (we orbit the sun, which goes around the galaxy, which itself moves through the cosmos)—get only distant glimpses of. Variations in color, light, heat, gravity, and elemental composition leave us with more questions than answers.

Meanwhile, do we really believe that we are the only intelligent, conscious beings in the entire cosmos? (What a waste of space!) Surely there must be other worlds in the creation besides our own with intelligent life, right? And, outside of the rebellion in heaven (Revelation 12), which shows that sin arose somewhere else in the creation and not here on earth, we have no reason to believe that God has sent a rescue mission to anywhere else besides the earth. Unfallen beings in other worlds likely interact directly with their Creator, as did Adam and Eve in the Garden of Eden before the fall. We see evidence that some of those beings were able to travel the cosmos, or at least to heaven, the abode of God (see Job 2:1). From what we could tell by the texts, these beings were able to remain in God's presence without succumbing to the power of His glory. Mankind, sinful mankind, would be destroyed merely by God's presence. Thus, part of what salvation and redemption ultimately accomplishes is eradicating sin so that we can live in His presence without fear of destruction.

The Hosts of Heaven

The point is that angels must have existed for long ages before the creation of the world. Psalm 8:5 says that "You have made him a little lower than the angels." Having lived for thousands of years, perhaps (if not more)—angels, unfallen angels, had plenty of time to acquire knowledge; and, not facing the mental degradation of sin, they surely exist at a much higher level of intelligence and ability than does sinful mankind. The creation account tells us nothing specifically about the creation of the angels, though we can infer that they were not included among what God has created during the six days of creation but were likely created at a separate creation sometime in the past.

Cherubs are also mentioned in Scripture, although with less frequency than are angels. Cherubs appear to have a guardian role. They are mentioned as guarding the entrance to the Garden of Eden in Genesis 3:24. They are seen covering the ark of the covenant, thus guarding the law of God (Exodus 25:20). They appear to be guarding the throne of God as well (See Psalm 99:1; Isaiah 37:16). In Ezekiel 10, one cherub takes fiery coals from the throne and hands it to a man clothed in linen, who scatters them over Jerusalem—perhaps a symbol of judgment on the holy city. Each cherub has four faces (a man, a lion, an ox, and an eagle), representing perhaps different parts of the creation: humanity, beasts, domestic animals, and birds. They appear to be involved in the protector role of watching over the events here as well as executing God's will on earth.

Lucifer was a cherub who deceived a third of the angels in heaven (Revelation 12), and then brought his rebellion to earth. Cherubs, winged like angels, are likely a subset of the greater angelic host, but with the defined role of guarding and protecting things such as the Garden of Eden, God's law, His throne, and His people.

Other beings

Job 1:6 reads: "Now there was a day when the sons of God came to present themselves before the LORD, and Satan also came among them." Where did this happen and who were these "sons of God"? Many Old Universe theorists believe that these are beings from other worlds, like angels, whom God had created long in the past. We know that other beings exist out there in the cosmos. Colossians 1:16 says, "For by Him all things were created that are in heaven and that are on earth, visible and invisible, whether thrones or dominions or principalities or powers. All things were created through Him and for Him." These other beings are the "invisible" things of His creation, the "principalities and powers" in heavenly places. It is likely that they had been created long ago, and not after the original rebellion in heaven that came to earth.

God's Moral Law

Psalm 119:142 says, "Your righteousness is an everlasting righteousness, and Your law is truth." God's law must surely have preceded the creation of this earth, or even of any of His intelligent moral beings anywhere else in the cosmos; otherwise, how would they know what is right or wrong, moral or immoral? How exactly God's law, the foundation of His government, is manifested in other parts of the creation (are there neighbor's donkeys to covet, or parents to honor if there are no progeny, as here on earth?)—we don't know. The principles espoused in the law of God are not temporary; they are eternal. If, for example, the angels are not the offspring of parents, the fifth commandment to honor one's parents must be extrapolated to mean to honor one's creator or predecessor. Show respect for authority. Again, we don't know how the law is manifested elsewhere, but it's hard to imagine created moral beings, beings who require freedom in order to be moral (a computer program might filter out pornography, but it's certainly not moral) not having a moral code to live by.

God's Natural Laws

The Hydroplate Theory, meanwhile, implies that God also has natural laws. He is not bound by the laws of physics and chemistry, but rather He works within them and uses them to affect His purposes. God, as Creator of these laws, can set them aside as He wishes. His creative power allows, and even necessitates, Him doing this at times. It's what we call a "miracle." For instance, how could God create something from nothing unless He exists "above" and transcendent to the natural laws that He had first created—and also sustains. Yet, when we look at the flood, the earth, the solar system, at the motion of objects, at weather, and at the functions of the human body—we can see the working of a God who also uses natural laws in order to further His purposes. Working backwards from these laws, we can to some degree extrapolate back to His initial creation of this planet, and to some of the natural laws that guided this creation after it was made, and even how, later, the global flood occurred. We believe the OUHP best describes the universe according to God's eternal natural laws.

Free Will

Another implication of an Old Universe is that of free will. The only way Lucifer and the angels could have rebelled against God is if they had free will from the start. God would not create the angels in order to purposely have them rebel against His ways. As stated above, in the context of God's law, if intelligent beings in other parts of the cosmos were moral beings, they had to have free will, free choice. This would include, obviously, humanity too.

The key issue here is love. Love, by definition, requires the option not to love. It requires moral freedom. God can force compliance but not love. Love that is forced is not love; it's something else. An Old Universe suggests that God's creation has always been afforded freedom of choice. This is a key element in the character and government of God. Whether in heaven, on other worlds, or on earth, wherever God has created moral beings, He has given them free will, free choice, and so the Calvinist idea that God predestined some people, no matter their free choices, to be lost is truly—besides being non-biblical—a horrible thought that makes God look bad, arbitrary, and tragically unfair.

Rebellion Confined

Scripture suggests that God sent His Son to this world, and that His sacrifice was unique to this planet, and this planet alone. Apart from the rebellion in heaven, there is no suggestion that sin exists in any of the other worlds that the Lord has made. (Again, we really can't think that we are the only planet with intelligent life, can we?) As we gaze out into the cosmos, we are looking in the direction of worlds without sin, even if we cannot see them now. Of course, powerful physical and chemical reactions are unfolding across the universe that we, from our very limited vantage point, can observe. However, there are activities in the cosmos that we cannot even detect, much less understand, and only when Jesus returns will so many questions be answered that leave us baffled now.

Humanity has a unique position in the cosmos: that the God of the universe would condescend to send His Son to this planet, this planet alone—*and die for it!* What a powerful expression of the love of God.

God's Unfathomable Sacrifice

Indeed, it is an amazing thought that God would sacrifice His Son for this tiny portion of creation, this tiny planet in the Milky Way, just one of two trillion galaxies. If the YEC view were correct, and this earth contained the only intelligent beings in the cosmos beside angels—then one could argue that God would, maybe even logically, should, condescend to send His Son to live and die, potentially an eternal death, to try to rescue the only intelligent and moral beings whom He created in the entire universe.

In contrast, if the universe is old, and if God had been in the business of creating for billions of years, and if He had invested time, effort, energy, and love in creating all the galaxies in His universe, many filled with intelligent life—then we can look at the cross as an even more amazing expression of His love. Amid a universe filled with life, God, out of love for the beings on this mere "blip"—*did what?*

He came down, incarnated into a human body, that of Jesus of Nazareth, lived among us and then, rather than eradicating this rebellious planet, which He easily could have done—He died for it, instead. Yes, the God who created and now sustains the entire cosmos, which is so big that we cannot comprehend it—He is the One who shrank down, became part of this tiny portion of His creation, *and then died for it.*

How do we, who cannot grasp the creation itself, grasp the One who created it, who stands above and beyond it and yet who holds not only all the stars in His hands but also our very breath? And then how do we grasp the idea that He died for us as well? What else matters if you believe this? Every other idea shrinks into insignificance before it.

How, then, do we respond to such amazing love, such an amazing condescension on our behalf? What could we do in order to adequately react to what we have been given? Anything and everything that we could do would seem so small, so trivial, so meaningless and empty before it.

Everything that is, except, perhaps . . . worship.

ENDNOTES

Chapter 1

1. Helge Kragh, "How Did the Big Bang Get Its Name? Here's the Real Story," *Nature*, March 25, 2024, https://www.nature.com/articles/d41586-024-00894-z.

2. Georges Lemaître, "The Beginning of the World from the Point of View of Quantum Theory," *Nature* 127 (May 9, 1931): 706, https://doi.org/10.1038/127706b0.

3. Fritz Zwicky, "Die Rotverschiebung von extragalaktischen Nebeln," *Helvetica Physica Acta* 6 (1933): 110–127.

4. Eric J. Lerner, "Evidence for a Non-Expanding Universe: Surface Brightness Data From HUDF," arXiv preprint, accessed May 26, 2025, https://arxiv.org/abs/astro-ph/0509611; Jayant Vishnu Narlikar, Geoffrey Burbidge, and R. G. Vishwakarma, "Cosmology and Cosmogony in a Cyclic Universe," *arXiv* preprint, published January 18, 2008, accessed May 26, 2025, https://arxiv.org/abs/0801.2965.

5. Ibid., 2.

6. Thomas J. Barrett, James F. J. Bryson, and Kalotina Geraki, "The Source of Hydrogen in Earth's Building Blocks," *Icarus* 436 (August 2025): 116588, https://doi.org/10.1016/j.icarus.2025.116588.

7. Eliza Richardson, "Ocean Origins," *Geosciences LibreTexts*, accessed May 26, 2025, https://geo.libretexts.org; "Carbonaceous Chondrite," ScienceDirect Topics, accessed May 26, 2025, https://www.sciencedirect.com/topics/earth-and-planetary-sci-

ences/carbonaceous-chondrite; R. Terik Daly and Peter H. Schultz, "The Delivery of Water by Impacts from Planetary Accretion to Present," *Science Advances* 4, no. 4 (April 25, 2018): eaar2632, https://doi.org/10.1126/sciadv.aar2632.

8. Steven K. Croft, "A First-Order Estimate of Shock Heating and Vaporization in Oceanic Impacts," in *Geological Implications of Impacts of Large Asteroids and Comets on the Earth,* ed. Leon T. Silver and Peter H. Schultz, Geological Society of America Special Paper 190 (Boulder, CO: Geological Society of America, 1982), 143–153.

9. "How Old Is the Sun?" *NASA Space Place,* accessed May 26, 2025, https://spaceplace.nasa.gov/sun-age/en/.

10. Tobias Owen, James E. Cess, and V. Ramanathan, "Enhanced CO_2 Greenhouse Effect to Compensate for Reduced Solar Luminosity on Early Earth," *Nature* 277 (February 8, 1979): 640–641.

11. Aleksandr Ivanovich Oparin, *The Origin of Life* (New York: Macmillan, 1924).

12. Francis Harry Compton Crick and Leslie Eleazer Orgel, "Directed Panspermia," *Icarus* 19, no. 3 (1973): 341–346.

13. John Burdon Sanderson Haldane, "The Origin of Life," The Rationalist Annual (1929); and Jack William Szostak, "The Narrow Road to the Deep Past: In Search of the Chemistry of the Origin of Life," *Angewandte Chemie International Edition* 56, no. 37 (2017): 11037–11043, https://doi.org/10.1002/anie.201703627.

14. António Vieira, Helena Maria Rodrigues da Silva, and Maria Alexandra Santos, "The Ambivalent Role of Water at the Origins of Life," *Life* 10, no. 9 (September 2020): 174, https://doi.org/10.3390/life10090174.

15. Sarah Lewin, "New Equation Tallies Odds of Life Beginning," *Space.com*, published July 8, 2016, https://www.space.com/33225-life-begins-equation.html.

16. Charles Robert Darwin, *On the Origin of Species* (London: John Murray, 1859).

17. Stephen Jay Gould, *Wonderful Life* (New York: W. W. Norton, 1989).

18. Luis Walter Alvarez et al., "Extraterrestrial Cause for the Cretaceous–Tertiary Extinction," *Science* 208, no. 4448 (1980): 1095–1108.

19. Jens Esmark (1824) and Louis Agassiz (1830s) are credited with early proposals of ice ages; no single scientist is credited with identifying all five major ice ages recognized in modern geology.

20. Charles Robert Darwin, *The Descent of Man* (London: John Murray, 1871).

21. Richard Charles Lewontin, "Billions and Billions of Demons," *New York Review of Books,* January 9, 1997, https://www.nybooks.com/articles/1997/01/09/billions-and-billions-of-demons/.

22. Francis Stephen Collins, *The Language of God: A Scientist Presents Evidence for Belief* (New York: Free Press, 2006).

23. Kenneth Robert Miller, *Finding Darwin's God: A Scientist's Search for Common Ground Between God and Evolution* (New York: HarperCollins, 2000).

24. Alister Edgar McGrath, *Darwinism and the Divine: Evolutionary Thought and Natural Theology* (Oxford: Wiley-Blackwell, 2011).

25. William Lane Craig, *In Quest of the Historical Adam: A Biblical and Scientific Exploration* (Grand Rapids, MI: Eerdmans, 2021).

26. Hugh Norman Ross, *A Matter of Days: Resolving a Creation Controversy* (Colorado Springs, CO: NavPress, 2004).

27. Gleason Leonard Archer Jr., *The Genesis Debate: Three Views on the Days of Creation* (Grand Rapids, MI: Baker, 2000).

28. Norman Leo Geisler, *Knowing the Truth About Creation: How It Happened and What It Means for Us* (Eugene, OR: Harvest House, 2003).

29. Bernard Ramm, *The Christian View of Science and Scripture* (Grand Rapids, MI: Eerdmans, 1954).

30. Ronald Leslie Numbers, *The Creationists* (Cambridge, MA: Harvard University Press, 2006).

Chapter 2

1. Peter Ludwig Berger, The Sacred Canopy: Elements of a Sociological Theory of Religion (Garden City, NY: Doubleday, 1967).

Chapter 3

1. John Clement Whitcomb Jr. and Henry Madison Morris, *The Genesis Flood* (Phillipsburg, NJ: Presbyterian and Reformed Publishing, 1961).

2. Kent Edwin Hovind, "Explaining the Flood without the Canopy Theory," *Creation Today,* April 2012, accessed May 26, 2025, https://creationtoday.org.

3. John Richard Baumgardner, "Catastrophic Plate Tectonics: The Physics behind the Genesis Flood," *Proceedings of the Fifth International Conference on Creationism* (Pittsburgh, PA, 2003).

4. Walter Thomas Brown, *In the Beginning: Compelling Evidence for Creation and the Flood* (Phoenix, AZ: Center for Scientific Creation, 1989).

5. Robert "Bob" Enyart, *Real Science Radio,* KGOV.com, accessed May 26, 2025, https://kgov.com.

6. Bryan Nickel, "The Hydroplate Theory Overview," YouTube video, accessed May 26, 2025, https://www.youtube.com.

7. "2022 Hunga Tonga–Hunga Ha⊠apai Eruption and Tsunami," Wikipedia, updated December 6, 2025, accessed May 26, 2025, https://en.wikipedia.org/wiki/2022_Hunga_Tonga%E2%80%93Hunga_Ha%CA%BBapai_eruption_and_tsunami.

8. European Space Agency, "Deuterium-to-Hydrogen Ratio in the Solar System," published 2014, accessed May 26, 2025, https://sci.esa.int/s/8kK7Bo8.

9. László Boros, "Deuteronomics," *Laszloboros.com,* white paper, published 2019, accessed May 26, 2025, https://laszloboros.com.

10. Tao, "Slowing Time: Deuterium-Depleted Water (DDW) and Its Emerging Role in Mitigating Aging Processes at the Cellular Level," published September 9, 2025, accessed May 26, 2025.

11. Kamran Yavari, Mahdi Ashofteh, and Zahra Ghorbani, "Anti-Aging Effects of Deuterium-Depleted Water on Mn-Induced Toxicity in the Fibroblast Cell Model," *ABI* 2, no. 3 (2024): 146–154.

12. Jie Qu et al., "The Biological Impact of Deuterium and Therapeutic Potential of Deuterium-Depleted Water," *Frontiers in Pharmacology* 15 (2024), https://doi.org/10.3389/fphar.2024.1431204.

13. Hungarian research team, "Deuterium-Depleted Water as a Complementary Therapy in Cancer Treatment," Biomedicines, published April 4, 2025.

Chapter 4

1. "The Big Bang: History of Earth," Wikipedia, accessed May 26, 2025, https://en.wikipedia.org.

2. Michael Joseph Behe, *Darwin's Black Box* (New York: Free Press, 1996).

3. Alpha Phoenix, "Speed of Light," YouTube video, accessed May 26, 2025, https://www.youtube.com.

4. Danny Ray Faulkner, "The Speed of Light and Cosmology," *Answers Research Journal* 6 (2013): 279–284.

5. Larry Vardiman, Andrew A. Snelling, and Eugene F. Chaffin, eds., *RATE II: Radioisotopes and the Age of the Earth: Results of a Young-Earth Creationist Research Initiative,* vol. 2 (San Diego, CA: Institute for Creation Research and the Creation Research Society, 2005).

6. Donald Russell Humphreys, "Young Helium Diffusion Age of Zircons Supports Accelerated Nuclear Decay," published November 1, 2005, accessed May 26, 2025.

7. Brian Thomas, "Magnetic Field Data Confirm Creation Model," *ICR News,* published December 28, 2010, accessed May 26, 2025.

8. Ann Gibbons, "Calibrating the Mitochondrial Clock," *Science* 279 (1998): 28–29.

9. Shi Yan et al., "Y Chromosomes of 40% of Chinese Descend from Three Neolithic Super-Grandfathers," *PLoS ONE* 9, no. 8 (2014), https://doi.org/10.1371/journal.pone.0105691.

10. Walter Thomas Brown, *In the Beginning,* chap. "Origin of Earth's Radioactivity" (Phoenix, AZ: Center for Scientific Creation, 1995).

11. Viktor Fedorovich Zelensky, "Experimental Device of Cold HD-Fusion Energy," *Condensed Matter Nuclear Science* (2017).

12. Walter Thomas Brown, *In the Beginning,* chap. "Liquefaction: The Origin of Strata and Layered Fossils" (Phoenix, AZ: Center for Scientific Creation, 1995).

13. Bryan Nickel, "The Hydroplate Theory Overview: Plateaus," YouTube video, accessed May 26, 2025, https://www.youtube.com.

14. Bryan Nickel, "The Hydroplate Theory Overview: Limestone Formation," YouTube video, accessed May 26, 2025, https://www.youtube.com.

More Resources

To access supplemental materials, recommended reading, and ongoing insights, visit my website at:

https://watersofthedeep.com/

New content is added regularly, so check back often.